I0132077

Robert Carstairs

Human Nature in Rural India

Robert Carstairs

Human Nature in Rural India

ISBN/EAN: 9783337366681

Printed in Europe, USA, Canada, Australia, Japan

Cover: Foto ©berggeist007 / pixelio.de

More available books at **www.hansebooks.com**

HUMAN NATURE IN RURAL INDIA

BY

R. CARSTAIRS

BENGAL CIVIL SERVICE
AUTHOR OF 'BRITISH WORK IN INDIA'

WILLIAM BLACKWOOD AND SONS
EDINBURGH AND LONDON
MDCCCXCV

All Rights reserved

CONTENTS.

HUMAN NATURE IN RURAL INDIA.

DREAMS.

IN the eyes of most people there is a contrast between the dreamer and the practical man, and dreams are looked upon as the
very opposite of realities. When I look back
on my life, however, and try to sort out its
dreams from its realities, I find that the
process is not so easy. Life is a procession
of facts, marshalled in order of time, which,
as they pass the little point in the universe
where I am, are the realities of the present.
We grope for them before they reach us, in
vain efforts to know them while they are yet

in the future; and as they come, they move
on, and are memories of the past. What,
then, are the realities of life?—I who see, or
the swift-moving procession that is passing
me? What are our dreams? Our bodily
senses are limited by time, place, and all the
various checks placed by the Almighty on the
powers of man. I am in a cage, against whose
bars my bodily powers beat in vain. But
the mind can travel far—passing through the
bars, and, by the aid of memory and imagination, wandering freely over and taking possession of vast regions of the past, the future,
the distant, and the unknown, from which the
body is for ever locked out. From experience,
from books, and from men, I receive knowledge
and information of all kinds, which I assimi-
late and make part of myself. That part of
this which relates to the future and the unknown present I call my dreams, and I look
on my dreams as greater facts than what are
called realities. What are these but a yester-

day's newspaper, to be cast aside and forgotten ?
But dreams remain. They do not file up and
disappear. A man's dreams are the man, far
more than is his past career. They influence
his actions for good or evil, and, through him,
the world. Our dreams give us the goals
towards which we strain, the standards we
attempt to reach. By our dreams we feel out
into the infinite, and to direct our dreams
aright is looked upon as the noblest task of
man. Dreams are the causes, the moving
powers, and what we call facts are but their
symptoms.

Dreams are called by many names,—love,
religion, patriotism, honour, lust, passion,
avarice, ambition,—and they are the great
facts.

Every man—even the small, the dull, the
poor—has his dreams. There are single men
—like Newton, Columbus, Stephenson, Peter
the Hermit, or Napoleon—whose dreams have
moved the world. There are times when

a community or a nation has been seized by
a dream as by an epidemic. The greatest
material fact for a man is death—his removal
from the standpoint whence he views the
procession which files up and passes on; and
yet there are dreams of many kinds, good and
bad, for which ordinary sane men are willing
to accept the risk, and sometimes even the
certainty, of death.

Dreams bring men into harmony, and
dreams also bring them into collision. They
make and sustain mighty works; they gather
into one vast number of forces: and then,
again, they shatter force against force, one
dream destroying the work that has been
built up by another. Dreams command the
forces of the world.

In these modern days the influence of
dreams is greater, because wider; more be-
neficent, because more powerful,—but more
dangerous. They are dangerous for the same
reasons that make the rule of a despot danger-

ous. Granted that the despot means well, still he must control affairs of which he knows very little. His own life, his own interests even, are not bound up in those affairs; and yet, by a nod, and, as it were, a word over his shoulder, he may all-unconsciously deal woe and destruction to thousands; or his ears may be closed to the cries of distress by the shouting of the few immediately about him. Then, again, men have a way of joining themselves into clubs or societies. They hold meetings, and repeat to one another their own arguments; they read the papers in which those arguments are echoed, and seldom look at or listen to the other side. Now when two interests draw apart, each absorbed in its own dream, and are not agreed, the end must be strife. So we find strife on every side : in industry, strikes and locks-out ; in commerce, tariff wars ; in the less civilised states, actual fighting ; and over the stronger states, the threatening thunder-clouds of war.

The cause of this danger is the isolation of dreams. They grow and grow, and, each becoming too great and proud to give way for the sake of harmony, will yield to nothing but force. In the relations between Great Britain and India, the danger that springs from isolation of dreams is greatly to be feared. However it came about, Britain is mistress of India, and when she speaks there is no one to say her nay. The dream that guides her actions is the dream of the ordinary British voter,—a good fellow enough, and full of common-sense, but knowing little and caring less about India and her affairs. His very anxiety to do right leads him to meddle, and, meddling without knowledge, he may unintentionally do much harm.

Now the best way to meet this danger is to bring home to Britain the dreams of India. Efforts are being constantly made to do this by natives of India who go to England, and by natives of Britain who have been to India.

I have no desire to disparage what others
have done. It has occurred to me, however,
that if I could record, not my experiences,
which are of little interest, but my dreams—
the influences which have moved me to action
from time to time—it might help my countrymen to understand Indian dreams a little
better. So far as my dreams are peculiar to
myself, they matter little; but as I have been
born and bred a Briton neither much better
nor much worse than the average of my
countrymen, my dreams are probably much
the same as those of my countrymen would
have been in the same conditions. In this
way, perhaps, my twenty years' experience
may be of use as a telescope to bring within
reach of the British eye objects otherwise
invisible.

HOME DREAMS

HOME DREAMS.

I DO not think that, until I actually started for India, I had any dreams about it.
I had read Macaulay's Essays, and heard
the hymn about "India's coral strand." I
believed all Bramins were priests, and that
snakes and tigers were met with everywhere. I understood the country was hot,
unhealthy, and mysterious; but was ready
enough to follow where so many had already
gone. I went out to India with ordinary
British feelings—prejudices, if you will—and
with goodwill in my heart towards the natives of the country.

Dreams come from trouble and want; and
my first dreams after leaving home arose from

absence. The native of Britain is more ready
than he of France to leave his home and go
abroad; but that, in my case at least, was
owing to our national custom, and not to
indifference. My stay-at-home countryman,
you to whom home and country and all that
they mean are a presence all your life
through, I wonder if it is possible for you
to feel for your home and country all that
the exile feels. You are not called on to
give them up, or even in these days to
make sacrifices for them. Your life is for
the most part taken up with struggles about
the management of the noble inheritance left
you by our fathers, and for a share of its
good things. But when a young man goes
away from all this,—when the faces he has
loved and known pass from his sight, and
all that he has known and that has been as
a part of his life is suddenly wrenched away,
and flung to a distance,—he may laugh and
seem cheerful, but his heart is sore.

The sickness of absence from home and country sometimes ends in death and forgetfulness, and the affections are transferred to new objects. In my case, however, as I believe in that of most of my countrymen, it ended in a double dream—love of home and love of country; and these two dreams have influenced me all my life through, restraining me and urging me on for the honour of home and country. If religion is the power of the love of God, then are love of home and country His good angels. And sorely is their help needed in times of sickness, of hard work, of hopes deferred, of tedious waiting, of bodily discomfort, of temptation, weakness, and worry, which make up the greater part of most lives, but above all, I think, of the lives of Britons in India. There are some who think that these dreams of home and country—to many all of home and country they are destined to possess in this life—should be laid aside by a man who is to do

real good to India. He should, say they, fix his
affections on India and things Indian alone.
They may be right for all I know, but I
have never found these dreams any hindrance
to my goodwill for India and the Indians,
and I very much doubt if the best way to
kindle love for another country is to kill
that of your own.

Then another dream that arose and grew
as I sailed eastwards was the greatness of
England. The ears at home are full of the
din of contending factions. There is not
an institution or a great man but has a
rival whose business it is to disparage the
other side; and this universal custom of
mud-throwing tends to destroy one's sense
of reverence and appreciation of greatness.
But when we got to a distance, all these
petty noises died out; and the mud that
was thrown disappeared from view, like the
weather - stains on a great statue. There
arose before my mind's eye the majestic

figure of Britannia, with her great representative the Queen, and her emblem flag.
As we saw her ships on every sea, and
began to realise how she sends forth her
armies of pioneers, merchants and teachers,
to all parts of the world, we were seized
with this dream of her greatness. It was
not her wealth or her wisdom, her ships or
her great possessions, that I thought the
most of. It was her *men*. The dream of
England's greatness awoke that other dream
which had always been there—the pride of
race—the feeling that I was one of a strong
and masterful race. Whether this dream of
race pride is one to be encouraged or discour-
aged there seems to be controversy; but it is
in the blood, and cannot be destroyed. If my
countrymen wish the sons of England to go
forth into the world without this pride of
race, then the rising generation must get a
very different training from those that went
before. There must be no more telling of

B

the noble deeds of our ancestors, of their bravery and heroism; for verily we go forth laden in the sight of our own country and of the world with a weight of prestige which is hard to carry save with the help of a proud heart.

I now began to attach a different meaning to that theory of the equality of men which, in common with nearly all my countrymen at home, I had been taught to believe.

At home I was in a British community, where natives of India, and of all foreign countries, were in a very small minority. Hospitality and curiosity combined to procure for these foreigners from the British communities in which they were living more attention than they would have got had they been Britons. They were accepted, being guests, not merely on an equal footing, but even as lions. But then they took no part in the management of affairs, nor did they compete with the British for employment in Britain. When competition arises the British race quickly alters its tone, whether with Polish Jews in London, or with China- men in America or Australia. It is possible to go through the polite forms of society with people of other races, and keep up the fiction, and even the sincere feeling, of equality, so long as you have not to do business with them, or to depend on them in some matter of vital importance; but we got to see, long before we reached India, that there were many reasons why we could not do business with, and depend on, men of other races as if they were our own people.

Then, again, as we approached India, it became more and more apparent that we were going among peoples who were broken up by great varieties among themselves. There were several of us young men to- gether on board ship, on our way to various parts of India. One was going to the Pun- jaub, among the Sikhs and Rajpoots; another

to the North - West Provinces, among the
Mussulmans and Hindoos of the plains; another to Bombay, among the Mahrattas; and
another to Madras, among the Hindoos of
the South. My own lot was cast in Lower
Bengal. We belonged to a strong race, and
how could each of us drop into a comfortable
equality with whatever race he might happen
to be thrown among? We were not made
of such chameleon-like materials. The Briton
is notoriously stubborn, holding by his own
customs and his own standards, and unable
readily to take the colour of his surroundings.
I know little about the others, but as for
me, I remained British. I could not, were
I willing, cast off British ways and feelings
or my character as a Briton among the fee-
bler races of India. In the lonely regions of
India—lonely to a Briton, if swarming with
men of other races—our dream of the greatness and strength of the British race is shared
by the vast majority of the people, and the

white man is hailed as the representative
of the strength, wisdom, and just dealing
of England. In the strength of this prestige
the English man or woman can travel alone,
unarmed, unguarded, through most parts of
India, in greater safety than in the heart of
England. In the strength of this prestige
the solitary constable will interpose between
hostile bands of fierce rioters, and they will
obey him and refrain from fighting. In the
faith of this prestige the postal runner, carrying her Majesty's mails, will boldly pass the
lair of the man-eating tiger by night, confident that even wild beasts dare not in-
terrupt the passage of the Queen's letter-
bags.

Now this prestige, or, as I have called it,
this dream of greatness and of strength, which
animates and supports every Briton in India,
which gives stoutness to the heart and power
to the arm of every native in the service of
the Government, is the prestige not of Lahore

or Delhi, Calcutta or Bombay, but of Britain. We Britons in India are the garrison, the outposts, the pioneers, the agents, the representatives of Britain. Cut us off from Britain; break, if you can, all our old feelings and associations; teach us to forget, if you can, our race pride, and get us to settle down as sons of the soil wherever we may be. You may spoil us for Britons, but even then you cannot turn us into natives. We shall but lose the best qualities of our race, and strengthen the bad.

There is another way that has been spoken of by which we should be assimilated to the people of the land. If we cannot grow like them, might they not be made to grow like us? To speak my mind honestly, I say, On the surface—yes; through and through—no. The people of the land I live among are the Bengalis. There are Bengalis—all honour to them—who are in advance of their fellows, but so few are our opportunities of testing

them to the full that even these I cannot
trust deeply. Courteous and friendly when
all is well; not very satisfactory to do business with at any time; in time of trouble and
adversity—broken reeds. That is what they
seem to me.

That thought—the time of trouble—makes
it hard for a Briton to accept a native of
Bengal as equivalent to a Briton. Supposing
the troubles of the Mutiny were to return, I
cannot, as a native of the country could and
probably would do, slip off my official dress,
and disappear like a drop in the ocean of the
native population. I should have to fight
with my back to the wall, and so would
my fellow-countrymen. We British in India
are no more always thinking of another Mutiny than our countrymen at home are always
thinking of a French invasion. Yet at home
the thought of a French invasion is a dream
that affects the national policy. They won't
have a Channel tunnel; they watch closely

the French navy. My lot in India has for the most part fallen in lonely places, remote from troops; far away even from the big cities, with their large European communities and ships. A very small local rising would suffice to finish off the lonely European, or group of three or four Europeans, who would be as much cut off from help as if they were on board a mutinous ship at sea. And if it came to that, the difference between a European and a native of Bengal would come out.

Then there is the social point of view. I did not understand this at home, and I fancy few Europeans at home do. When a native of India goes to England, he generally lays aside his Indian habits — lives in rooms built on the European pattern, eats European food, and sits down to meals with Christians. Some of these, when they return to India, retain European ways, but many of them do not; and the vast mass of the

natives of India have wholly different ways.
If it be granted that, for the purposes of
official work, the native of Bengal is equal
to the European, this accounts for only a
portion of the day—six or eight hours out
of the twenty-four. I am an official, liable
to sudden transfer, and may succeed, or be
succeeded by, another official at very short
notice. If a European succeeds a European,
it is comparatively plain sailing. The successor will probably take over the house and
the bulk of the furniture, and socially there
will be little change. But if a native succeeds a European, he may not care for the
house, preferring to live in the bazar; so the
house is not taken over, the native ideas of
a dwelling-house varying considerably from
the European. Then the native would probably not care for the furniture, and that
would have either to be sold at a heavy loss,
or carried off, also at great cost. The de-
parting European has probably been a mem-

ber of a mutton club; but the native does not eat mutton, so that is broken up. The ladies of the native's house probably do not go out, or see company, and here is another loss. Where the European society consists of two or three families, the substitution for one of them of a native family disorganises everything. If, on the other hand, a European succeeds a native, he most probably finds no proper house or furniture, and nothing as, in his eyes, it should be.

There is one more obstacle to the notion of equality, and that is the inconvenient feelings aroused by race pride — a pride begotten not in India but in Britain. From top to bottom, Britishers hate being ordered about or reprimanded by foreigners; and if they find a difficulty in standing it from

> "A Rooshian,
> A Frenchman, Turk, or Prooshian,
> Or perhaps Italian,"

they are not likely to stand it from a

"Bengali,
Marathi, or Nepáli,
Or perhaps Mahomedan."

This feeling of taking no orders from any but
one's own countrymen is in the heart of all
Britishers in India—kept under in the better
educated, but burning more and more fiercely
among the less polished and more natural
rank and file.   It is fiercest of all probably
among those of mixed race, who cling more
desperately to the European strain in their
blood the more remote it is.

This dream of race pride, strong and in-
eradicable, which has possessed me ever since
I landed in India, is natural, and general,
and must continue if we are to hold up the
standard of British prestige.   There are agita-
tors and faddists who are trying to scoff it
away, but they will not be able ; and to beat
it down, but they will fail.

Of the great dream of all—religion—I will

say little here. I was brought up in the
faith of my fathers, and never let it go. I
can only say that its influence on my life has
been always for good; and my regret is that
the weakness of human nature did not suffer
me to live more nearly up to the perfect
standard it held up for me.

These, then, are the four great dreams that
I brought with me from home—love of home,
love of country, race pride, and religion.

A FLEETING LIFE

A FLEETING LIFE.

THE voyage itself was the beginning of a new life, and, dream-like in itself, made the beginning of another dream—the fleeting nature of our life in India. We meet and part. We have to hold everything lightly —life, home, friends, property—ready to let all go when the time comes. At home, we used to hear the same thing from the pulpit, and many wise injunctions on the head of it. To the young all things look permanent. It is when we grow older that time and all things human begin to seem unstable. But life in India drives home the lesson severely, and with little delay.

It was a pleasant time, that first voyage, but a dream.

> " Joyfully, joyfully plough we the ocean,
> Through the bright drops of light flitting in motion;
> Seize, as the minutes fly, wingèd with pleasure,
> Seize and make captive hours laden with treasure !
>
> Merrily, merrily lighting the waves up,
> Bright flashing phosphorus grey ocean laves up ;
> Floating on, quickly gone, such is our pleasure ;
> Seize and make captive, then, Time's passing treasure !
>
> Fearfully, tearfully speak we of parting ;
> At the old memories tear-drops are starting ;
> Maidens sing murmuring, mourning lost pleasures,
> While Time is on the wing, bearing his treasures !
>
> Lo the sun, though his last radiance he's throwing,
> Sends his light, golden bright, o'er the sea glowing !
> Minutes fly, speeding by ; each has its pleasure :
> Think of these, quickly seize Time's passing treasure ! "

Ah ! the mood was then hope, and we had youth and health, and the buoyancy they give. Some fortunate ones have been able to retain this mood all their life through, carrying the " sunlight of their presence " into many a gloomy little corner of the great kaleidoscope of Anglo-Indian society. But the prevailing effect of this dream when it is continued too long is a feeling of unrest, of discontent, and, long before it ends, of weariness to end it. Think what it means, you who live quietly at home. We go to you now and then, those of us who can afford it, and, ten years after, we see everything the same as it was before. You are most likely living in the same house, and there we see the faces that we knew long ago—the old a little older, only the young changed. There is the same old furniture, the garden, the old familiar landmarks, the same neighbours and tradesmen; the same pew in church, and the same old family doctor. Those whose faces we once knew, and whom we see no more, are lying in the graveyard with their ancestors, and you tend their graves with affection. How can you realise our Indian life?

I am far away from my older friends, in a strange place, where perhaps in the course of a year or two I have become familiar with my house and its neighbourhood, and formed friendships with a few of my neighbours, European and native. I am getting to know my furniture and servants, and to take an interest in the garden. All at once a telegram comes, and I have to make out a list of all I have—furniture, stores, horses, &c.—except my clothes, a few books, and some portable treasures; sell them for what they will fetch; be off, perhaps within three days of getting orders, and drop down a couple of hundred miles away, among a new set of people; make new friends, and gather new possessions in a new house. If I am as long as three years in one place, probably I see the neighbours changed several times over. Even in times of sickness, when one wants to be quiet, we cannot always rest, for we are ordered off to sea or to the hills. India is not an easy country to get well in.

We know how Britons at home look on their friendship. It is a gift, not to be lightly bestowed, nor to be hastily withdrawn. If a stranger settles in a neighbourhood, he will perhaps take several years to win the confidence of the neighbours, which, once given, remains for generations. So particular are they, that many worthy people cannot get into their society at all.

We who have left England to live in India see something of this spirit — enough to remind us that it still lives in our own country — when we visit home occasionally, and on board ship among the tourists, or "globe - trotters" as they are called. In India we see little of it, even among the tourists, for they accept frankly and genially the hospitality of total strangers for weeks or months at a time. If any of these strangers presumes on this kindly intercourse

to hope for its renewal at home, he is most
often disappointed, and does not always
hide his feelings. Doubtless, however, such
feelings are unreasonable. If an Englishman
does at Rome as the Romans do, that is no
good reason for expecting him to do the
same ever after, even if it leaves him owing
debts which he can repay in no other way.
If the Englishman with a home, an estate,
a large income, a good position in society,
and plenty of leisure, falls in with the only
way which renders life tolerable in our ever-
changing society of comparatively poor, home-
less, and hard-worked people, it is the only
thing he can do, unless he stays away from
it altogether. The fault is theirs who exert
themselves beyond what they can afford for
his entertainment.

In India we cannot take time to form
acquaintances. One result of this constant
change is to destroy the British reserve,
and make every one of us behave to his

neighbours like an old friend for the time.
When the time comes we part, and see and
hear no more of one another till the changes
of life bring us together again. We form
friendships—dear and enduring—which, tried
by the most severe tests, do not fail. But
they do not depend on daily intercourse and
bodily presence.

The very frequency of the changes made
among European officials supplies a remedy
for the hunger of the soul which they produce; for if "the world is a small place,"
still smaller is the Anglo-Indian world; and
if we part, it is most often to meet again
at some other time and place. Every one
has a reputation which in time penetrates
to all parts of the province; and when he
drops into a new neighbourhood, his neighbours know him, and he his neighbours,
though they may never have met before. This
is one reason why Anglo-Indians who meet
as strangers invariably begin to talk of people,

owing to which custom reputations are quickly spread, and the process of settling down
becomes easier.

One relation there is which is an exception
to the constant changes, and that is the relation between husband and wife. The old
saying, " Where thou goest I will go," has
a meaning to us wanderers which it can never
have to my stay-at-home countrymen. In
his wife a man has a comrade who shares
his life wherever he may go, and whom he
can never be required by a telegram to drop
and leave.

But this relation also brings with it that
which weans us more than anything from
the love of this life in India, and that is
the parting from wife and family for the
sake of their health, and the proper education
of the children. In my own life I have had
some of this, but not so much as many I
know. I know a man—himself an invalid—
whose wife and children are at home in Eng-

land.   His wife because of her health, his
children because of their age, cannot be with
him.   He lives on his mail letters, and goes
on, from week to week, from month to month,
from year to year, looking for the time when
he may retire on his pension and join them.

The plains of India are no permanent home
for Europeans.   Even merchants and planters,
whose home is more settled than that of
officials, are unable to look on this country
as their home in the same way that our
brethren in Australia, Canada, and the Cape
can.   For that, we all look away across the
sea : we are "strangers and pilgrims."

INDIA

INDIA.

I BELIEVE a great many people at home think of India as of a place about as big as a county in England. Of course they know it is not, but that is how they think of it. They lump it all together into one, and when a rising takes place in Assam, Bill says to Tom, "Them Hafgans is at it again." I remember a friend who came out to see India wrote to me from Allahabad, some 450 miles off, that he had been asking about me there. " I knew you were somewhere in these parts," he said. Various causes hide the distances and masses of human life in India. There are the single Government and Law everywhere, the English language and English ways in all

the chief towns and cities; the railways and
facilities for travel; the post-office and telegraph, which convey letters and messages all
over India as cheaply as is done within England alone. Then we are not in England
impressed by mere areas and distances, being
familiar with the much greater areas and
distances of America, Siberia, Africa, and
Australia. The map of India used in England
is on a small scale, showing perhaps less than
one in a thousand of her towns and villages. In short, Englishmen look on India
as a man looks on a cheese, not seeing the
mites.

As one lives on in India, the feeling arises
that it is easier to make acquaintance with all
Europe than with India as a whole. Before
we have lived long in the country, and while
our imagination has not been dulled by too
much contact with actual facts, we are perhaps
better qualified than afterwards to form broad
views and express them with confidence. A

certain bold disregard of detail is almost
necessary if a strong effect is wanted. I feel
that a twenty years' training of close attention
to detail has to a great extent deprived me of
any power in that way. I shall attempt rather
to describe the effect on my own mind of this
long personal contact with India and its people,
than to impress lessons or information on the
minds of others. Not only are impressions on
my own mind facts which I am entitled to
claim some knowledge of, but by confining
myself to them I strengthen my position in
two ways. One of these is, that I come forward as a witness, not as a judge, claiming
nothing more than a belief in my sincerity ;
the other is, that I do not profess a complete
knowledge of the subject, but admit that my
knowledge is limited and partial. My career
has been in Lower Bengal, one out of the
many provinces of India, and even within
that province the greater portion is personally
unknown to me. It is of the people I have

seen and lived among that I would be understood to speak.

I see them every day of my life. They surround and haunt me. I cannot get away from them. I have got used to their hovering about my room—peeping round the corner, entering unannounced. They pour along the roads, swarm in the bazars; start up wherever I may go, even in the fields and the jungles. And yet I do not see them; for in my presence they behave differently from what they do out of it, and I have to guess—to dream— how they act and speak when free from its restraint. In my presence they are most often acting a part. If I overhear a conversation it is most likely intended to reach my ear. As an official, I am a marked man, and a busy man. Where I am is generally known, as also where I am likely to be, so that stage effects are not difficult to produce. I am not a Haroun al Raschid, to go about among the people in disguise, and the approach of a white

face and European clothes at once puts every
one on his guard.

When a Frenchman wishes to study the
nature of the English, how does he set about
it? If he means to be thorough, he will first
get introductions to families, and study them
in their homes, living with them, eating and
conversing with them. Or perhaps he will
study their newspapers, their novels and literature. Most Frenchmen draw on their imagination, and dream. I cannot tell how a
conscientious Frenchman who wished to learn
the English character would manage, if de-
barred from social intercourse, and from information through literature. That, however,
is my position with regard to my Indian
neighbours. Their customs forbid social intercourse, and their literature is, with a few
exceptions, not informing on the point.

Such information as I have had has been
gained chiefly from litigation and from talks.
I have had before me during my twenty years

of Indian life scores of thousands of law cases
civil and criminal, and these have given me
many a peep into the social and family life of
the country, otherwise veiled from me. True,
when there is a lawsuit, that means that something has gone wrong, and the constant study
of lawsuits is apt to give one a morbid view of
the ordinary life of the native. True also, at
such a time there is much lying and fraud, and
concealment of the real facts. But, after
making full allowance for these, a law case is
a valuable opportunity, since it makes necessary the close study of a single set of facts
and a single group of people, in circumstances
where the law allows little or no reserve. It
is the nearest we Europeans can get to the
inner life of the native. Again, the native of
India has little imagination, and even his lies
are instructive, showing what he is accustomed to see and think about. Get him
to talk, and he consciously or unconsciously
gives information about himself, which, as

we gain experience, we are able to test
and sift.

Then I have spent, besides the longer time
when I lived at my headquarters, some 1500
days in touring about among the villages,
during which time I have traversed some
15,000 miles, going with my tents from village to village, mostly over ground unknown
to the tourist, unheard of by Europeans
in the greater Indian towns. On these
rounds I have seen and conversed with men
of all degrees, rich and poor, great and small,
learned and ignorant, and of all religions,
castes, and occupations. I have talked with
them on their own affairs, and heard their
wants stated in their own words. I have
discussed with them the various troubles that
they suffered from, and the possible remedies, and endeavoured, with what success I
myself can hardly judge, to see things from
their point of view, which is often very different from the official one. I have had reason

D

to believe that, different as is my race and remote my home, I possessed the confidence, and, I would fain think, also the affection of many among them. I dare not say I know them, but I can honestly say that all this time I have tried hard to understand them. I have had dreams on dreams about them, and some of these I propose to put down here.

HODGE

HODGE.

MY native neighbours are divided roughly
into three classes—agriculturists; those
who live on them, giving service in return;
and those who live on them without giving
service in return. It is our custom to talk
more of the non-agricultural classes than of
the agricultural, and to be severe on those
who, in our opinion, oppress and fleece the
latter. I cannot help thinking that the
community must be taken as a whole. Here
are the oppressors and there the victims,
who are a hundred to one of the oppressors.
If they let those whom we are accustomed
to regard as their oppressors take money or
property or service from them, it may be

under sheer compulsion; but it may also be because they are getting in return some service which we do not appreciate but which they think adequate. India, again, is bound by custom, and, as every one knows, customs established in certain circumstances may have been good at the time, though the circumstances have since changed.

I wish to concentrate interest for the present, not on the parasites, but on the creature which supports them; and therefore, putting away the castes and races which figure most in the eye of the public and of England,— the Bramin or priest; the *kayest* or writer; the *bunniah* or trader: putting away also all dwellers in towns, and even the great communities of the boatmen, carters, weavers, potters, oilmen, and others,—I turn the eye of imagination on the vast agricultural class, which, were the country blest with universal suffrage, would swamp over and over again all the rest put together. How does neigh-

bour Hodge live? What does he think of?
What are his hopes, fears, ambitions? How
(most important of all) does he quarrel, and
what about?

As my mind ranges over the regions I
have known,—Eastern Bengal; the suburbs
of Calcutta; the undulating plains of Midnapore; the alluvial river-banks and irrigated
soil of Arrah, and the hills and valleys of
the Sonthal country,—the first thought that
arises is, that Hodge in Bengal is various.
I have seen only a few of the many Districts
of Bengal, and in those Districts a few only
of the many millions who inhabit them; and
yet I have come across many varieties of
Hodge. How many more, then, must there
be as yet unseen? The main varieties of
Hodge are the fighting man, the drudge,
and the pioneer. The fighting men are to
be found chiefly on the banks of the great
rivers,—the Ganges, the Bramaputra, and
the Megna,—where, after each year's floods,

thousands of acres of fertile land are found
to have been formed or swept away, according
to the caprice of the stream. The Rajpoots,
or warrior Hindoos, and the Firazis, or Puritan Mahomedans,—men who do not feast or
dance or smile,—are on the watch to rush in,
and be the first to occupy the new land, holding it against all comers by the strong arm.
The drudges are found in those vast expanses
of plain that have been for generations under
close cultivation. They live where their
fathers lived, and cultivate the same fields,
forming, as they increase in numbers, " con-
gested districts," where the soil groans be-
neath the population it has to support. These
are the great mass of agriculturists, both
Hindoo and Mussulman. The pioneers are
found in the less thickly settled country,
where there is yet jungle to clear and waste
land to reclaim—on the borders of Eastern
Bengal, in the Terai, in the delta of the
Ganges, and in the less dense but less fertile

forests of South-Western Bengal. While not refusing to fight with men, they rather wage war with wild beasts and with nature.

I have spoken of the pugnacious varieties of Hodge,—such as the Rajpoots and Ahirs of Behar; the gloomy and stern Mussulman Puritans of the lower Ganges; the sturdy Chandals of Fureedpore; the wild Bhuyas, Kols, and Sonthals of the south-west,—and of the cowardly varieties, of whom are the mass of the Hindoo cultivators of the plain. When we talk of fighting men, however, it must be with reserve; for of all these varieties not one is looked upon as fit to furnish good material for the native army, which is recruited from outside Bengal. Courage and cowardice are taken as relative, not absolute terms, and serve to measure one Hodge in the province against another. Hodge does not travel, and is gregarious. Like draws to like, associates and quarrels with like. There are 150,000 villages in the province

I live in, mostly occupied by varieties of
Hodge; and each of these villages is made
up of one or more separate parts, each part
being occupied as a rule by a single variety
of Hodge. It is the people of the same
quarter who come in contact with one another. We do not ordinarily find Rajpoot
fighting with Mussulman, or Mussulman with
Chandal, because they seldom come in contact—their interests do not clash. It is most
often Rajpoot who quarrels with Rajpoot,
and Chandal with Chandal. Cowardice
matched with cowardice is not afraid to
fight, and thus it comes to pass that desperate
quarrels go on, not only in the villages inhabited by the plucky varieties of Hodge,
but also in those inhabited by the cowardly
varieties.

Hodge has a very narrow horizon, generally
bounded by the village and the market-town.
He thinks, not of the wide world, but of his
own little world—the village. I remember

once, when we were out driving, the ponies
were being changed, and by some accident
both got loose. They scampered off in great
spirits, and seemed in for a spree. The
whole country was open to them to roam
over. But no! nothing would serve them
but to fly at one another, and they went
on kicking and biting until they were parted
and caught. So it is with the villager in
Bengal. His interest in outside matters is
almost nothing, and why? Because his
mind is held firm by village interests. His
dreams are within the village, and seldom wander elsewhere; and his rage, when
kindled, burns, not against the distant
stranger, but against his kinsmen and
neighbours.

VILLAGE LIFE

VILLAGE LIFE.

THE highest class of the village community are the peasant proprietors, who practically own the land they occupy, paying rent to the landlord, and being under certain restrictions as to use and alienation. Next to them are what we would call the farmers, who rent land from the permanent owners. And last, the labourers, who work for wages in cash, or, more often, in kind. The houses of the village are gathered in groups, according to the number of quarters, and round the hamlet lie the fields. The fields of a villager are not all in a ring-fence, but are mixed up with the fields of others. The only marks of division are the

little ridges which are made round the rice-
fields to retain water in the rainy season.
There are seldom any fences. Hodge builds
and repairs his own house; provides his own
manure, seed, tools, and cattle; and sells his
own produce.

What sort of a life do they live in this
miniature world? It is a matter whose interest is not to be measured by the material
wealth involved, because there is room in
this small world for tragedy. The community in which Cain killed Abel was neither
wealthy nor large. The main interest of
Hodge, as of all mankind, is how to make
a living, and in an agricultural community
the things men talk and think about are
land, the weather, crops, cattle, and markets.
Hodge is, we will say, one of the peasant
proprietors of the village. He cultivates a
few acres of land, out of which he has to
make enough to live on, and to meet his
expenses. He is too proud, as a rule, to

work for hire.   In Eastern Bengal especially,
this is so much the case, that labour for the
roads and for public works has generally to
be imported from the west.   I remember
once when there was a flood in the river,
which was 8 feet above the level of the
country, only kept back from devastating
it by a rotten embankment.   The embankment was all but topped, and was showing
signs of a flaw, and we were at our wits'
end for labour to stop the breach.   Just
below the place where a breach threatened
was a village, which must certainly be swept
away first if the embankment broke, lives
being endangered and property destroyed.
But not a soul in the village would stir.
They said this was coolies' work, and the
business of the Government, and they would
do no work except on their own fields.
They felt about it just as a London clerk
would feel if called on to work as a blacksmith.

E

Hodge is various in the skill which he brings to bear on his land. Everything runs in castes in India, and there are castes of famous cultivators,—the Koiris, for example, who have a name for the cultivation of vegetables; and the Barois, who devote themselves to *pán*, that vine whose leaf is valued by all Hindoos as a stimulant. Others, again,—the great majority,—content themselves with the standard crops—rice, maize, pulses, jute, and the like, with a patch of potatoes or sugar-cane thrown in. Others again, of whom the Sonthals are most prominent, are mighty pioneers, carrying out great engineering works to turn a ravine into fertile rice-fields, but have no enterprise and little industry as cultivators, weeding little, sowing late, and cultivating no crops to speak of save the standard ones of rice and maize.

When Hodge cannot cultivate his land himself, he sublets part of it, taking rent in cash or kind ; or hires labourers to cultivate it, whose wages are usually paid in kind. This is how the lesser fry of the village— the small men who have not enough of land, or none at all—are provided for. Those who have capital get land to farm where they can, trusting, on a vacancy, to rise into the higher ranks of the village society. There are rises and falls in life. A large family with many strong men and much land is suddenly re- duced by cholera or fever, and there are not left in it hands enough to till the family fields. Or a man grows poor through trouble or misconduct. Again, a man has a large family of sons who grow up into strong workers, wanting more land. These waxings and wanings, the changes they bring, their anticipation, their happening, and the study of how to profit by them, are one great field of village politics. Every head of a household has to think of the time when his children will grow up, and will need room ; so there is no such thing as content. Each one presses

to enlarge his holding. In England, Hodge's
children, if they can get nothing to do at
home, can enlist, or go to sea, or emigrate,
or go to the towns. Labour that is willing
can generally gain a living. For utter failure
there is charity or the poorhouse. But Hodge
in Bengal can look for no such relief. The
army, the sea, the colony, or the town is not
for him. He has no charity or poorhouse to
look to. By the land he must live; for the
land he hungers, and to the land he is fast
moored. What little emigration to new lands
is still possible gives scant relief to the con-
gested millions of peasantry. Hodge in the
mass works out his destiny where he is.

Hodge is head of a household, yet his may
be only one of several households in the homestead; for the custom is that, when the sons
grow up and marry and have children, all the
families remain united in one, eating together,
cultivating their joint fields, having their
property in common. Very pretty, this

harmony, while it lasts; but it does not always last. The close contact and divided duties frequently give rise to jealousies and quarrels; and quarrels between those who are in contact every day and all day, and who cannot get out of sight of one another, are apt to become very bitter indeed. Hodge has to watch anxiously and act warily to avoid the risk of such a misfortune.

Independent of such complications, Hodge has his domestic anxieties. If Hodge has grown-up sons who are married, they bring their wives to live in the house. The governess of the household is the mother, who keeps the younger wives to their tasks of cooking, fetching water, scrubbing dishes, husking rice, sweeping out the house, washing clothes, gathering sticks, and the many miscellaneous duties that have to be done. The women do not, as a rule, work in the fields, go to market, fodder the cattle, spin or weave. Very old men and very young

boys and girls herd the cattle. If there is
a widow or a child-wife, the heavy part of
the household drudgery is laid on her. In
a household there are numberless occasions
for quarrelling between the members, — because the passions of jealousy, ambition,
rivalry, and envy in their narrow minds
are just as easily roused by trifles as they
are in more cultivated minds by things of
more importance. With neighbours it is
the same. The life of Hodge is an open-
air life. He has, indeed, room for the shelter
of man and beast, but most of the household
life is spent in the open air of the courtyard.
Neighbours can hear all that is said in neighbours' houses; children will quarrel, animals
will stray, and misunderstandings arise from
many causes.

There are also the more delicate social
troubles which exist wherever men live.
There are no unmarried women, but many
unhappy, young, and foolish wives, and

many widows. Village scandals, true or false, are constantly flying about, and much of Hodge's attention is needed for guarding the fair fame of his house. There is also the " black sheep"—the brother, or cousin, or son who has become idle, vicious, or criminal. In this land, where the ties and the importance of blood are so much thought of, the honest and deserving members of a family suffer more in reputation for the sin of its erring ones than in countries where the individual is more isolated. Moreover, the men and women of a family who have gone wrong do not so often disappear— though sometimes they do—from the family horizon, as in England. They remain in the neighbourhood, a thorn in the side of the family, and especially of Hodge, its head.

Hodge and his neighbours have many points of contact in the fields, even where there is no dispute about land going on. He has

practical matter to watch. There are his
boundaries. Fields are usually separated
by ridges of earth ; and if the fields on opposite sides of a ridge belong to different
men, a keen watch has to be kept lest the
boundary line, the middle of the ridge, should
be shifted over by the neighbour shaving
a slice off his own side. Then there is right
of way. In most villages there are no
proper roads, and Hodge reaches his field
by threading his way along the boundary
ridges of his neighbours. When men do
this, it matters little ; but when cattle pass,
crops are damaged, and many are the debates
and minute the etiquette by which quarrels
on this subject are settled or prevented.

Then cattle and goats will stray. In a
land where there are no fences, scanty pasture, and many cattle and goats (usually
not less than one to an acre), very great
care in the herding is necessary to prevent
damage to standing crops. But the cattle

are for the most part in charge of little
children and very old men. Children will
play, and old men will doze. It sounds
pretty to read of the lass who

"Left a' her flock to stray
Amang the bonny bloomin' heather;"

but when flocks and herds stray among standing crops, the case is different. The owner
of the field or one of his family discovers
the trespass, and there is a hullabaloo, with
much vigorous language, unless the owner
is one of those powerful men whose cattle
may graze where they like, none gainsaying.
Many are the debates among the village
elders owing to the straying of cattle, and
much diplomacy is needed to avoid ill-feeling
or defiance; for war means retaliation, and
Hodge is business man enough to know that
it would be a poor consolation for his own
trampled and wasted crops to see those of
his neighbour suffering as much.

Another knotty problem for Hodge is how

to get wood. He needs fuel, unless he
has given up hope of wood, and taken to
dried cow - dung cakes instead. He needs
also posts and rafters for his house, and
ploughs. In England there is always the
market, and no resource but the market.
English Hodge has his house provided for
him, and does not bother about posts or
rafters. He gets the use of a plough to
work with; and where he does not use coal
for fuel, brushwood is easily got. But in
India there is no market, and in most villages
there are few trees. Hodge must get wood,
and get it where he can. Little care is
taken of forests, and the land is being
denuded, so that in some places there is
not enough of wood to supply the ordinary
needs of the people. This not only causes
great anxiety, but demoralises many men
otherwise respectable, who have come to look
on wood-theft as many men in England do
on poaching, and used to do on smuggling,

with a sneaking feeling that it is not a
regular crime.

Then the cattle. The fodder difficulty is
always present. In Bengal no hay is made.
There are no turnips or root crops for cattle,
and silos have not yet taken hold. From
October to June very little rain falls. With
the spread of cultivation, the forest and waste
land has been much reduced, and nearly all
land that will bear herbage is under crop.
In Bengal, meat not being an ordinary article
of food, pasture is not laid down. The cattle,
when the crop is on the ground, are often
penned up at home. When and where there
is no crop they are taken out in herds, with
the goats, to pick up the best living they can
from the scanty herbage. They do moder-
ately well, with the help of a shower or two,
up to February, and then the country begins
to dry up in earnest. Sometimes there falls
a shower, bringing up a green flush, but more
often there is little or nothing to eat that one

can see ; and when the rain comes in May or June, and ploughing begins, the cattle are mere bags of skin and bone, weak and unfit for work. It is difficult to see how this is to be avoided, with a human population of over 500 to the square mile, and cattle, sheep, and goats more numerous than the human population. Such land as will bear crop (and land that will not bear crop is not much good for grass) is all being brought under cultivation. Hodge in the mass is opposed to the destruction of the village pasture, but the individual Hodge, who always wishes to extend his holding, is often tempted to nibble at the pasture, and sometimes to swallow up a lump of it. Once destroyed, it is seldom restored.

Then there is the cattle disease, which, owing to religious prejudices, cannot be checked as in England by the slaughter of infected animals. Neither do men understand it, nor

will they incur trouble or expense to check it.
I have seen a cow-keeper with sixty cows who
had rinderpest among them, and was urged at
least to separate the sick from the sound, but
would not, because "it was too much trouble
to feed them in two lots"! Hodge must have
cattle ready for the ploughing season when it
comes round, and the problem how to do this
gives him much grave thought.

Then there are the great questions of irrigation and drainage, which may lie dormant
year after year, but become raging in a season
of drought or flood. In an ordinary year, the
water store for irrigation is enough for all,
the supply being abundant, and less of it
being needed. But when the rain fails, and
the crops are dying, it becomes a matter of
life and death for each man in the village to
get hold of as many of the last few precious
drops as he can for his thirsty fields, when all
would not suffice. Then, if they can stave off

actual violence, the old men of the village argue and bargain and arrange so that there may be as much as possible for the crops, and it may be fairly divided, while a sufficient water-supply is reserved for the drinking of men and cattle, and to keep the fish alive.

INFLUENCES

INFLUENCES.

HODGE has his neighbours and kinsmen, with whom he has relations more or less intimate. His kinsmen are ranged in two circles—an inner circle of relations, and an outer of the caste. Seldom does Hodge know, or in a social sense care for, any one outside his caste. The Mussulman in theory knows no caste, but with him also exclusive dealing is the rule,—artificial castes, or what they call messes (those who will eat together), being with him the limit of social intercourse. Each caste has its own customs even in such an important and general matter as marriage. When it is realised that Hodge in India always belongs to some caste or

mess, interested in the affairs of no other
caste or mess ; that there are hundreds of
such societies, and each of these is broken
up into hundreds of local branches—neighbourhood being isolated from neighbourhood ;
and that castes and messes are not like the
old Scottish clans, each contained in its own
glen, but scattered and mixed up with one
another,—the great difference in conditions
between the life of Hodge in England and
that of Hodge in Bengal will become clearer.

In Bengal a Mussulman worshipping one
God, who is a Spirit, lives in the same village
with a Hindoo, who keeps an idol in his
house. The women go to fetch water from
the same tank; the children play together
in the village street; the men shoulder their
ploughs, and march out side by side to the
same fields, and there perhaps work within
a few yards of one another season after
season ; watch together their ripe crops by
night; go to market together; pay rent to

the same landlord ; and sit together in the same *punchayet* (village council) to determine a dispute among neighbours. But they may not eat together, or intermarry, and their lines of social life are as far apart as if they were cows and sheep respectively.

A man who does not follow his caste is not respectable, either in the eyes of his neighbours or in his own eyes. The faith of one who has given up his caste is like an uprooted plant which, unless it finds fresh and congenial soil—sometimes even then—withers up and dies. Hodge in social matters is strongly under two influences—that of the women and that of the caste.

In all societies women form half the population, and the children, who are, cutting a different way, also half, are under their influence. Hodge is a marrying man. To be respectable and succeed in life, he must marry and have children. I remember attending a meeting of advanced natives on the subject of the emancipation of women, where one youthful speaker waxed very eloquent on the wrongs and sufferings of the women,—mewed up in zenanas; kept like slaves; and not allowed to have any freedom or any opinions. He was pleading for the opening of the zenana. After he had sat down, an old man got up and said: "My young friend has spoken loud words here, saying that our women are slaves, who have no freedom. I would like to see him go and repeat all this at home. The fact is, we are all hen-pecked." And the roar of applause that followed this frank avowal showed that the speaker had hit the nail on the head. An old friend of mine, a native, who had successfully, as chairman, managed what was long known as a model municipality, once told me: "When I want to get anything done, I go to the women and try to get them round. It is no use attempting anything

if they are against it." Who has not heard
of zenana influence?

The influence of woman over Hodge is
made of several strands. The first and
strongest, as all the world over, is love.
Hodge is a domestic man, and loves his
wife and children. There are exceptions, of
course, when husband and wife cannot pull
together; but in the main, the family gets
on quietly and harmoniously. The quarrels
and failures, being conspicuous and attract-
ing notice, bulk largely in the public eye,
but are few in comparison with the whole
mass of married life. In married life, Hindoo
Hodge compares favourably with Mussulman
Hodge, whose loose divorce laws and meaner
estimate of women lessen his respect for
his wife.

The next strand is his dependence on his
wife. She cooks for him, and manages
indoors all his affairs. Here Mussulman
Hodge feels the influence too, and an alarm-

ing one it is in some circumstances. When, in Eastern Bengal, marriage registration for Mussulmans was being introduced, it was at first unpopular, and the reason assigned was the fear lest, if the knot should be too tightly tied, many wives might poison their husbands.

Again, the wife keeps the secret hoard that in most households is reserved for extremities, and all the household treasures. Who so deeply interested in the welfare of the home as the wife, its unseen heart? It is pathetic in my eyes, when there has been a burglary, and the wife is called as a witness to identify the property recovered. Here is a quilt, which she follows round and round the edge till she comes to a tear which she had sewn up. There is a brass pot with the familiar dent on the lip. A woman was giving evidence once, and the judge cast doubts on her identification, because there were no special marks. "I use these things

every day, and handle them. Blindfold me
and I will pick them out!" They blind-
folded her, and she did pick out all her
things from among the others on the table.
Well, the husband knows his wife's concentrated attention and devotion to the interests
of the home, and greatly is he influenced by
her advice in all his transactions. He may
often look only to his present ease or pleas-
ure; but she looks to the interest of her
children, as all true mothers do.

Probably the wife's influence over Hodge
is the stronger because he sees so little of
other women. They do not go to market,
or work in the fields. If Hodge goes to
a party at a friend's house, the women may
have cooked the food, but do not appear.
The food is served by men, and men only
sit down. The women associate together,
going to the tank for water, and in their
merry-makings. Even in the same household there are strong restraints on inter-

course, the younger brother's wife not daring to speak to her husband's elder brother. The effect of all this is no doubt to strengthen greatly the influence of the two women with whom a man is on terms of perfectly free intercourse, the mother and the wife.

The influence of the women is generally conservative. They hold strongly by old religious beliefs, old traditions, and old customs, and have a horror of what is new. This, I have little doubt, intensifies the resistance offered by Hodge to innovations,—what we call reforms, — even if they are meant for his benefit. He deeply, and, I must confess, too often with justice, distrusts them, as they do not always lead to the results hoped for. There is much that is to him unknown,—beyond the reach of his imagination even,—and he is so much accustomed to seeing that those who make such proposals have axes of their own to grind, that he takes refuge in the customs of his fathers as the one tried guide who will not be found wanting.

As for his children, Hodge spoils them. They are under no discipline, and do as they like, which is all the worse for them when they are grown up.

The next influence that affects Hodge strongly is that of society. Hodge leans heavily on the narrow circle of relatives and caste-fellows in which he moves. It becomes his conscience.

In England, outside his social circle, Hodge is subject to the influence of " the squire, the parson of the parish, and the attorney"; the minister and the schoolmaster; the newspaper; the union or friendly society; the public-house—the poor man's club—the franchise, the proceedings of the law courts, and all local governing authorities; the sea, the railway, the road, the post-office; cities and city life; the wanderings and tales of those of

his neighbours and friends who are soldiers,
sailors, emigrants, or in employment at a
distance. All these neutralise to a greater
or less extent the influence of the little
clique of his own folks to which Hodge belongs. Imagine them all removed, and you
have Hodge in Bengal—a small peasant proprietor, unhappy if alone, and dependent,
outside his family, on caste for society.
Religion I will come to later.

Many people praise the caste system; and
undoubtedly, as practically the only force in
rural India that binds men together, it is
valuable and useful. Its weak point, however, is that it does not provide a stable and
lofty standard of conduct. The standard is
a fluctuating one, depending on the disposition of the members for the time being.
If that is for evil, caste works evil and not
good. To illustrate the limitations of caste
(or among the Mussulmans, the mess), I will
note three incidents which I came across in

the course of my duty (I do not say that India is the only country where such things can be found).

One of the best of the Bramins in Scrampore, which abounds in Bramins, was talking with a missionary, and lamenting the prevalence of beef - eating and liquor - drinking among their young men. "Why," said the missionary, "do you not put your caste rules in motion and stop it?" He replied that each family was as bad as the others. If a man denounced the offence in another family, the other family would retaliate on his, and both families would then be put to the expense of purging their uncleanness. So they preferred to wink at it all round.

In a certain low caste of Hindoos, in which remarriage of widows and deserted wives is allowed, a strong party arose in favour of allowing a woman to leave her husband and marry another when she chose, practically abolishing marriage. The older and better members lamented this, but said there were so many against them, what could they do?

In a Mussulman village, whose inhabitants nearly all made a living as pedlars, there was faction. The leader of one faction was given to the use of *ganja*, an intoxicating drug. So the leaders of the other faction decided to lay a trap for him. They called a meeting of the village together, and said : "Let us all become very holy. Let us all sign a covenant that we will keep the Koran and use no intoxicating drugs." They hoped that the *ganja*-eater would refuse to sign, and lose influence in the village. But he met them on their own ground, and said : "I am ready to sign, but let us in our covenant include the whole Koran and not a part only. Let us enter a clause that we will take no interest on money." "But that is what we live by; how can we do that?" said the others. "I can't help that. It is forbidden by the Koran as well as the use of intoxicating drugs," said he. So there was a good deal of discussion and bargaining, and as a result a covenant was drawn up which contained no mention of intoxicating drugs or of interest on money, and which all were able cheerfully to sign.

These and numerous other instances that I have seen convince me that society, however strong its influence, offers to Hodge no steady guidance. It cannot be better than what it is elsewhere — the average of its members.

Hodge's social duties are mainly those connected with births, marriages, and deaths. The caste has certain servants,—the barber, washerman, &c. If the caste is displeased with Hodge, he is, as we should now say, boycotted. His invitations are not accepted, nor does he get any. Men may not give him or take from him food or water; he cannot get wives for his sons or husbands for his daughters, nor can he get the services of the caste servants. In short, he and his are "left severely alone" until he has made his peace, usually by turning away the offender from his family, or by the payment of a fine and giving a feast. Hodge, as I have said, is gregarious, and the isolation is an intolerable burden to him, which he is fain to get rid of at any cost. This he does, if the caste is unanimous, by submission; if it is not, factions may be formed.

Like all societies, the caste has its leaders; and the leader of a society formed of ignorant men, as this is, will probably become a "boss." By "boss" I mean one having at his command a body of men whose duty it is to obey him without question. It is the "boss" who generally pronounces whether a man is to be outcasted or not. Sometimes there are rivals for the office of "boss," and one of these will wish to put a man out while the other tries to keep him in. Then factions are formed. Hodge, if not himself a "boss,"

is bound by the influence of his caste or
faction "boss," and his great study is to
avoid being denounced. In Bengal the com-
mandment, "Thou shalt not be denounced
to the caste," corresponds with that in England, "Thou shalt not be found out."

As regards religion, I have never been able
to make up my mind entirely. A religion
cannot be safely judged by the conduct of
its professors, for then a Hindoo, for instance,
would have a very false notion of our religion.
I do not therefore feel qualified to gauge the
exact nature of the influence on Hodge which
is exerted by his religion. Comparing his
feelings with our own two great command-
ments, "Thou shalt love the Lord thy
God," and "Thou shalt love thy neighbour
as thyself," my notion is that Hodge fears
rather than loves his God or gods; that he
does for his neighbours nothing more than
social duty requires him (and that is, outside
the caste, nothing); but that, reading "rela-

tives" for "neighbours," he does more for
them than we do. He is not taught to give,
unless it be for feeding Bramins or preserving
cows. His religion seems to be looked on by
him as a mechanical one, of works rather
than of the spirit, in which a man may gain
merit or bring on himself punishment by acts
of which he was not conscious.

The Bramin priest is a social servant,
who is necessary at such ceremonies as the
naming of a child, or marriage, or religious
services, rather than a teacher. The mulla
is much the same. There are holy men
who go about holding what may be likened
to revival meetings, and arousing religious excitements; but there is nothing corresponding to our parish clergy, who urge
and teach religious truth week by week.
Hodge in Bengal has no doubt his share
of natural religion; but in the main his religion is the caste or mess in which his lot
lies, and its opinion is his standard. As the

caste or mess rarely concerns itself with matters outside its own circle, its opinion is generally narrow, concentrated, and, on all high
or distant problems, blank. Where, however,
any one can succeed in interesting all or many
of these small societies in any matter, such
as was done in the case of the kine-killing
agitation, no power can be imagined more
formidable in India. The number of men
affected is enormous, and it is difficult to
gauge the depth of their enthusiasm or the
lengths to which it will lead them.

HEALTH

HEALTH.

TO Hodge, as to all men, a very important question is that of health, and this comes under two main heads — sanitation and medical treatment.

As regards sanitation, that is usually left to nature. There are whole districts which are pest-ridden, the haunt of malaria, cholera, and other plagues. The unfortunate people who cannot get away from these, live on in listless submission to their fate. Hodge, as a rule, thinks very little of endeavouring by the art of man to lessen the terrors of a sickly climate. The village site having been chosen in as suitable a place as can be found, — high and dry, with a fairly well - cleared country round, and a water-supply within reach,—sanitary arrangements are not further extended. The village roads get worn down into deep hollows, full of ruts, and reeking with mud and filth. The premises of the various dwellings are also well stocked with rubbish and dirt of all sorts. In Bengal especially, ventilation is prevented by the thickets of trees and bamboos.

The water, that most important of all necessaries, is got, where there is no river or running stream, from wells or from tanks —open artificial excavations filled with rain and surface water. It was formerly an act of virtue for a rich Hindoo to dig a tank; but, unfortunately, the same virtue does not seem to belong to repairing the old as to digging the new. What with the silting up of the old tanks with vegetable refuse; their pollution for agricultural purposes, as for steeping jute or hemp; their deliberate destruction in order to get a rood or two of rice laid for cultivation; and their depletion in dry years for irrigating the fields, — Hodge's water - supply is often very bad. What he has he makes bad use of, for he will bathe his cattle, bathe himself, and wash his dirty clothes in the reservoir itself, and then drink the nastiness afterwards. He will also throw the remains of the dead, especially of those who have died of infectious diseases or epidemics, like cholera, into or beside water, and even bury them in the bed of a stream or the side of a tank. Sometimes, when there is a proper water-supply, the women are too lazy to go and fetch it. I remember going into a village where there had been cholera, and trying to find out the cause. In the first house I was told there was a very fine drinking-water tank some quarter of a mile distant, and that the drinking-water came from there. But when I asked where they got their water

to cook with, they showed me the rice-field
at the back of the house, which was reeking
with putrefaction and rapidly drying up.
It was the same in the next house, except
that there the supply for cooking came from
a shallow pond of rotten water at the bottom of a deep pit, and covered with green
slime. Where the water was taken, there
the dishes were washed, and the very clothes
of those who had died of the cholera.

Hodge's utter recklessness on the subject of
defiling water probably arises from the fact that
he holds a wrong theory about it. Water is the
great purifier, and he believes that it purifies
everything, and is of so great virtue that it
cannot itself be defiled. Hence he sees no
necessity for keeping the purifier itself pure.
We have a proverb that virtue is its own
reward. Hodge seems to think that virtue is
its own protector, or at least that the mission
of protecting it has not been imposed on him.
There are men of such high caste that they

can do what they like, being in the same way considered incapable of impurity. If the cattle disease breaks out among the cows he worships, he leaves their protection to the gods; if impurities get into the water, that is the affair of the water. When a place is unhealthy, he will blame the air and the water, but do little or nothing to mend their defects.

Hodge's general belief is that his duty is to do his caste work, and nothing else. He is as particular as a trade-unionist about that. He accepts from nature, or from the Government and the philanthropic rich man, who are nature's instruments, all such gifts as good air, light, and water, and even such artificial things as roads and bridges. He never thinks of cleaning or levelling a road for himself, but will contentedly wade through the same miry puddle, or break his cart-wheel, generation after generation, over the same old boulder; or will sit patiently by the stream in flood until it subsides and he can ford it. It is all custom. He is used to his troubles, and, having a very small imagination, never thinks how it would be if they were removed. When the smooth road, the railway, or the bridge has been made, it is used freely by Hodge, and in a year or two you catch him grumbling over the roughness or obstacle— him who till so lately never dreamt of getting anything so good as he has. Custom again ! The new becomes the old, and a fresh point of departure.

It is the same with drainage. Hodge lives contentedly and unthinkingly by a marsh that is flooding his village with malaria, until some one thinks of draining it. His first feeling when such a work is proposed will probably be one of indignation at the removal of an old landmark, and the loss of his supply of prawns. Even when the health of the village improves, he will probably put that down to a freak of nature; and the only benefit he will acknowledge from the im-

provement is the addition to the amount of
cultivable land. I doubt if in this respect
Hodge in Bengal is so very different from
Hodge in England, where sanitation is a
modern science, and there is a proverb that
"dirt and cosiness" go together.

The open-air life, and the small size of the
villages, save Hodge in Bengal from many of
the worst sanitary evils that have to be com-
bated in the towns of Britain. As for the
towns of Bengal, they have been put through
a course of education by the British Government, which has removed many of the worst
of the old evils, most frequently in the teeth
of strong opposition, for all the world over
a man clings to nothing so closely as to his
nuisances.

We come now to the second head—viz.,
medical treatment. Hodge's diseases come
on him, as a rule, in two forms—one in the
shape of general climatic diseases, such as
malarious fever; and the other in the shape

of epidemics, such as cholera or smallpox. I am going to discuss, not the causes of these afflictions, which are many and the subject of much controversy, but how Hodge looks on them and meets them.

He looks on disease, not as a warning of nature to put right something that is wrong, but as a punishment from God. He submits to it patiently—tamely, we should say—accepting it as fate; and his main indigenous remedy is processions, prayers, and sacrifices, by which he hopes to propitiate the wrath of God.

When a sudden change of climate occurs—from a river, for instance, deserting its old bed—enormous destruction of life may take place. The most conspicuous instance I have heard of was the destruction of Gour. This city was the capital of Bengal, and stood on the banks of the river Ganges. The Ganges changed its course, and a terrible pestilence destroyed all the inhabitants who could not

get away. Since that time Gour has been
an uninhabited mass of ruins, buried in dense
jungle. In modern times there was the Burd-
wan fever, said to have been caused by the
blocking out by an embankment of the flood-
waters which used to wash large areas. This
swept over large and populous villages, leaving behind it a succession of jungle-grown
heaps of ruins.

When an epidemic breaks out there is a
panic for the time, and after it is over, careless
indifference. Frantic grasping at any remedy
when the evil is imminent, but no precau-
tions taken when they can be taken with
effect. There are, for the country districts,
no colleges or trained physicians, no drug-
shops or druggists.

In cases of ordinary disease, when Hodge
resorts to a physician at all, there is generally
some one in the neighbourhood—a man or an
old woman—with a hereditary reputation that
way, and a knowledge of herbs and simple

remedies, whom he calls in. Among these there are some, especially in the towns, who have considerable skill, and know things that our doctors are ignorant of; but the greater proportion are neither highly gifted nor well trained. The mysteries of physic they may know something of, but of surgery they know nothing. It is natural, therefore, that Hodge should not expect the physician or such drugs as he can command to do much good. Suffering is in abundance among those dense masses of humanity, where death is ever reaping a rich harvest. It has just to be borne.

Great confidence is felt in the white man and his medicine. In the medicine, apart from the man, the faith is little. This is probably owing to adulteration, and to keeping things that ought to be thrown away. But the faith in the man is touching sometimes. An old woman came to one of our hospitals and asked a friend of mine, " Can't you give

me something for my eyes? My children are
turning me out because I can't see." It was
old age that was making her blind. "You
will never see better," he said. "Oh, but you
must have some medicine to make me see!"
she moaned; "think again! My children are
turning me out!" I remember an incident
which shows how cheaply a reputation as a
medicine-man may be gained. I was travel-
ling through a village, and a man turned out
with a very bad leg, which he asked me to
prescribe for. I declined to prescribe, but
offered to pay for a cart to take him to the
hospital about fifteen miles off. He never
went; but for some time after people were
bringing me patients to heal, on the strength
of the wonderful cure I had made of this
man's leg!

Government hospitals are getting more
patients, but were and are still unpopular
with Hodge. The reasons for this are, for
one, that there is a nasty echo in the rooms,

which are larger than the little huts Hodge
is used to, and no lights at night. It is a
dreary thing for a sick man to wear through
the long night in the dark alone, or with
a coughing neighbour and an echo.

Then, the medicine is not always good,
and in towns at least the dispensary is picketed by the messengers of quacks, who inter-
cept patients and warn them that it is no
good going there. Then Hodge is impatient
for effects. He expects to swallow the medicine and be well straight away, so that he
will not always carry out a course of treatment.

Quacks are a growing nuisance. They
are distinct from native practitioners, for
they call themselves " Doctors," against which,
unfortunately, there is no law, and profess
to practise English medicine though without
qualification. Their way of making money
is low fees and long medicine bills. I knew
one village of about 2000 inhabitants in

which were sixteen quacks. Their existence is a reproach and a hindrance to English medicine, and prevents Hodge having that confidence in the science which he generally has in the man.

I do not know how far the belief in witchcraft extends. Many of those who call themselves Hindoos and Mussulmans have it, while in the minds of nearly all aboriginal people, such as the Sonthals and Kols, it is a cardinal article of faith. When a woman is reputed to be a witch, she is held responsible for all the sickness of the neighbourhood, whether of men or of cattle. If she is not put an end to or tortured herself, her husband is fined as a punishment, and husbands have been known to kill their own wives merely to escape from the constant bleeding of their purse by fines.

Thus Hodge has his various remedies for plague and sickness.

H

WAYS AND MEANS

WAYS AND MEANS.

ALL the world over, the budget is the chief matter of anxiety—how to make it balance. From the Chancellor of the Exchequer to the meanest head of a household, the same problem has to be solved, and if the scale be small, the balancing is quite as difficult.

Hodge has to trust for his ways and means to the produce of his fields. A few of the richer men do a little trade, or money-lending, or ply a cart; a few of the poorer go out as labourers. Hodge in the mass, however, lives on his fields. His income, varying with the area and quality of the land, is difficult to reduce to a money value, because so little of it is ever turned into money. I fancy, however, that any one who has, after paying rent and cost of cultivation, anything over £5 a-year to spend, is looked on as a fairly warm man, and there are some who have to keep their families on as little as £3 a-year, or even less.

Without going into statistics, I am trying to enter into Hodge's mind, and read there the anxious thoughts with which he studies his great little problem of fitting all his expenses into this small frame. Talk about cheese-paring! He has plenty of experience in that. There is the rent, and the necessary expense of cultivation—cattle for ploughing, seed, implements, and labour, if there is not enough in the house. Then there is his food, and that of his family. Any of his field produce used as food must come out of income ; and there are things he must buy, such as salt, oil, tobacco, and spices. Then there are clothes, for the women do

not, as a rule, spin or weave; and fuel, if he
cannot get it for nothing. The house has
to be kept in repair, with rafters, posts, and
thatch, which have to be frequently renewed.
There are the various social expenses,—the
fees to the barber and other village servants;
the costs of little festivals, at naming of the
child, weddings, and funerals; ornaments
and dowry at weddings; fees to the priest
and physician; the rent; presents to the
landlord's servants, interest on debt—caste
fines. There are few conjurers who can do
more with a hat than commonplace, everyday,
stupid Hodge has to do with that £5 a-year.

What does he do when he has a surplus?
That, unfortunately, is not a usual occurrence,
even where the personal expenditure leaves
a balance on the right side. When he has
money to spare, he sometimes invests it in
cattle, dishes, or clothes; sometimes in getting more land; sometimes in making a bid
for influence in the village, by litigation or

otherwise ; sometimes he lends it out at interest on mortgage of land or taking articles
in pawn ; sometimes he spends it in reduc-
ing his debts ; sometimes he hides it, to form
a secret hoard. The secret hoard is the reserve fund of the family, usually in charge of
the women, and such hoards are frequently
discovered, probably lost through the sudden
death by cholera or other calamity of all in
the secret. Occasionally there are signs of
a banking system, where a money - lender
takes money on deposit, paying interest, and
uses it in his business. Sometimes the surplus
is used to stock a small shop, or to set up
a cart and bullocks. Hodge in the mass is not
fond of litigation, but there are parts of the
country where that seems to be a favourite
way of getting rid of surplus wealth.

If there is a deficit, how is it met ? Hodge
has no savings bank, or friendly society, or
bank, in the English sense, to fall back upon.
He has four ways of meeting deficit—by re-

ducing (save the mark !) expenditure ; by fail-
ing to meet obligations ; by exhausting hoards
and selling property ; and by borrowing. Some
few enterprising members of the class go and
work in the service of others, doing occasionally what those who cannot live by their land
must do regularly ; but partly from pride, and
partly from want of work near home and shy-
ness about going far afield, Hodge does not
often indent on this resource. Indeed it is
not a practical one except it be habitual, because the time is not propitious for seeking to
earn money by labour when employers are becoming candidates for work, and most causes
of shortness affect a whole neighbourhood at
once. Hodge runs up bills at the shop, gets
into arrears with his landlord's rent, and with
the interest of his debt ; pawns, sells, mort-
gages, and borrows as far as he can, and sits
tight waiting for better times. If better times
do not come, Hodge and his family either starve
to death or sink into the ranks of the landless.

HODGE'S MASTERS

HODGE'S MASTERS.

THESE are three in number—viz., the landlord, the money-lender, and the village ruffian. I am now looking at them from the outside—from Hodge's point of view. The caste "boss" has already been discussed.

The landlord may be large or small, but he seldom lays out money on his estate. His budget, so far as the land goes, is one of receipts only. This is the general rule, to which there are, of course, exceptions.

The vast majority of landlords are small, though there are landlords who have large estates. From Hodge's point of view, the main question about the landlord is, whether he is contented or hard up; and the next,

whether he keeps a firm hand on his servants
and middlemen or not. Unfortunately for
Hodge, nearly every landlord is either himself
a money-lender turned landholder, in which
case he is eager to make much more, or in the
hands of money-lenders, and driven by them.
Again, the ordinary landlord is ignorant, surrounded from youth by adulation and indulg-
ence; idle, never going out or attending to
business; and leaves his affairs in the hands
of ill-paid, greedy, and unfaithful servants—
hands to which much of Hodge's money sticks
before reaching the master. To save himself
from losses of this kind, the landlord frequently alienates by lease for a term or for
ever all his rights to speculators, whose one
object is to make money by the investment.
Then a wise father may be followed by a
foolish son; or the landlord's family, where the
estate is not kept together by primogeniture
or some such rule, may not agree to enjoy the
property in a friendly way, and may quarrel

desperately over their rights. Or the landlord
may quarrel with a neighbour about a boundary, or, in short, a dozen things may happen
to make trouble, into which Hodge is dragged.

When the landlord or his servant wishes to
make money out of the estate, Hodge is the
bee that must produce the honey. As land is
made to yield its produce by ploughing and
harrowing, so is Hodge ploughed and harrowed
by a bad landlord and his greedy servants.
Woe to him who makes a display of wealth! It
draws the spoiler on him. How Hodge defends
himself we shall see further on. One thing
may be noticed, that the old-fashioned landlord
generally tries to get money out of his tenants by means of occasional contributions, and
the new-fashioned by means of higher rents.

The landlord renders Hodge services, and
does not merely take his means away. He
is the local centre of order, and settles disputes between village and village, keeping
up a sort of discipline too over the caste

"bosses." He is supposed to stand up for his tenants against outsiders, and to satisfy all the demands of Government against them. I say he is supposed to do this, although—partly from the incapacity of the landlords, and partly from the growing independence of the tenants—serious inroads on this principle have been made. The landlord also used to, but now does not often, keep large reserves of grain in case of famine, and occasionally, if inclined that way, introduces an improvement, such as a road, a drain, or an irrigation work. Hodge looks on such unusual outbreaks of benevolence with a suspicious eye, not wholly without reason. An old landlord once told me, at the time when the Tenancy Bill was being discussed, "They may make what laws they like, but we can always get the ryots to agree to our terms. There are ways. For instance, I can start an improvement such as a drain, and ruin a man's land." I am only repeating what I heard, and do not vouch for it. It serves to account for Hodge's suspicious attitude towards benevolent actions which an outsider would expect to call forth a gush of gratitude.

The money-lender's is the rival influence to that of the landlord; and as two stones, either of which singly would fall on a man and crush him, may, by supporting one another, give him a safe shelter, so the landlord and the money-lender between them often give him breathing-space.

In a land where capital is scarce, communications are difficult, and the scale on which Hodge does his business is so small, the banking system is necessarily very different from that usual in a rich country like England. In any part of the United Kingdom we have banks with agents, well paid and able, disposing of abundant capital, and financing at moderate interest the operations of the farmers. In the Bengal village we have the private

I

capitalist, with a very petty capital, and he deals with Hodge on very hard terms.

I think the extortions of the money-lender are sometimes described in exaggerated terms. The risks are considerable, and, because of the smallness of transactions, the working expenses are great. When a creditor has to call perhaps twenty times, making the journey of some miles on his pony, to collect a debt of a few shillings, the rate of interest needed to cover his expenses in time and trouble must be high.

Hodge is generally in debt to a money-lender. If he were got out of debt, he would probably lose no time in getting into it again. Why is this, if the money-lender is such an oppressor? It is chiefly because he feels the comfort of having a man with money interested in keeping him going. If he falls into trouble through litigation or failure of crop, the money-lender who has invested in him will spend what he can afford before he

will sacrifice the investment already made.
The investment means not money at so much
per cent, but simply that Hodge is in debt
and pays from time to time what he can
afford to give. The creditor keeps all the
accounts, which Hodge is not, as a rule, clever
enough to check. But he would indeed be a
clever man who could make much of the
accounts, for the transactions are more often
in kind than in money. It is not easy to add
up three bushels of grain, a kid, a pot of
clarified butter, eighteen annas in cash, two
pumpkins, and a bundle of straw. They rub
along somehow, Hodge giving what he can
afford and the money-lender taking what he
can get, with little extra presents that do not
go down in the accounts, such as bundles of
firewood, milk, and oil when visits are paid,
until something occurs to break the connection. But a connection of this kind often
lasts for generations.

The personal system, which means that the

banking business is carried out of the living
fund instead of out of savings, presses hard
on Hodge when his creditor dies, leaving several sons who have all with their families to
live out of a capital that did not much more
than suffice for one. Then the cords are
tightened, and Hodge is squeezed.

The third of Hodge's local masters is the
village ruffian. I am not now talking of that
nuisance the village thief, who lives by levy-
ing stealthily small contributions from his
neighbours' crops; nor that still greater nuisance the cattle-poisoner, who destroys cattle
for the sake of getting their skins. These are
looked on and treated as vermin, to be extir-
pated if possible. But the village ruffian is
the man who levies blackmail by virtue of his
power to do mischief. Do we not know something of this man nearer home through the
discussions on the Irish troubles? It is need-
less to enumerate the many ways in which the
village ruffian can exert his power. One will

suffice. Hodge lives in a thatched house with stacks of straw beside it, and at certain seasons a spark will make it blaze up. If nobody else understands what that means, Hodge does.

OFFICIALS

OFFICIALS.

THE officials whose influence is felt by Hodge are the local authorities, the police, the magistrate, and the Civil Court. I will pass quickly by the local authorities, some of whom have considerable powers of annoyance, but have little direct power over life and liberty.

The police, from Hodge's point of view, consist of the village watchman nearest, the Government police constable next; then the head constable or sub-inspector of the station, and the inspector of the division; and last, in the remote distance, the superintendent of the district. The more remote central authorities Hodge sees and thinks of no more than we see or think of any possible inhabitants of the sun. The highest official who may be said to have any contact with Hodge's world is the sub-inspector.

The British reader can hardly realise the position of the police in Bengal, nor can a British official, save with difficulty, put himself fairly in Hodge's place, and look at them from his—the outside—point of view. In England there are as many magistrates as policemen; in Bengal there are practically no local magistrates, and of all the magistrates in a large district (say with two millions of inhabitants) there is only one who has any control over the police. I except, of course, the village watchman, whom Hodge looks on rather as one of his own people than as one of the police. Imagine—the nearest approach to a parallel—the sheriff of a Scottish county left to control, with the aid of one chief constable, and without any of the magistrates who at present aid him, the police of an area as large

as the county of Inverness and as thickly
populated as the Lothians. That is one of
the many items of a District Magistrate's
duty in Bengal.

The village watchman is strictly local, paid
by the villagers, and never transferred from
his beat. The Government constable is strictly Imperial, paid by the central Government,
and moved about all over Bengal at the will
of the central authorities. On an enormously
larger scale, and much less strongly and
efficiently manned or trained, it is more like
the Royal Irish Constabulary than like anything in England, where each county has its
separate staff, control, and responsibility. No
local people, high or low, official or non-official,
with the one exception mentioned above, have
a word to say as to the pay, merits, disposition, or discipline of any policeman, officer or
private. They cannot get him appointed or
dismissed, punished or promoted. The constable's pay is slightly above that of the la-

bourer. There is no coroner; there are seldom to be found gentlemen of position,
medical men, clergymen, professional men, or
men of property and leisure; so that the
play can often be played in the dark.

The police station is seldom within five
miles,—often it is over ten, sometimes twenty
or more miles distant,—and from the village
it must be reached across country—through
mud or jungle, and across unbridged streams.

Hodge's general associations with the police
are connected with the phrase, "Shell out."
A policeman, wherever he goes, has free quarters and the best of everything. To enumerate all the powers of a police officer by which
he can serve or harm Hodge would be tedious.
It is easy to see, where there are no magistrates, men of position, coroners, medical men,
newspapers or other means of publicity, how
murders, suicides, and other occurrences can
be hushed up or false charges concocted. The
power of arrest alone, which every police officer

above the rank of constable has, is a formidable weapon. I shall never forget the face of a well-to-do and important man, who remained in hiding when there was a warrant out against him, and found his way to me at last to give himself up and be admitted to bail. He was a stout man, evidently in ordinary life bearing himself with the portly dignity of an alderman; but he came to me crouching, with cheeks flabby, pouched, and shaking like a jelly, and staring eyes. This was solely from dread of arrest.

Pretexts for arrest can be found very readily. I remember, in a big town, one of the richest men of the place was charged with gang-robbery, one of the most serious offences known to the law. The police, as they are bound to do when they arrest a man on such a charge, refused bail. The gang-robbery alleged was an assault on a beggar, and the property stolen was the worthless rag he wore on his shoulders. I

judge of the fear that natives have of arrest
by the efforts they will make, and the money
they will spend, simply to be let out on bail.
What the police do to inspire dread it is not
my purpose to attempt here to describe. No
doubt they commit illegalities or irregularities,
besides using their formidable legal powers, for
illegalities are often necessary to get things
done. The European landlord who determined to run his estate on Christian principles — *i.e.*, without employing clubmen —
found it impossible. " The captain cut off
my hair as a punishment," said a soldier to
his comrade. " He can't do it : it's agin the
law," was the answer. " But he did," said
Tommy. " But I tell you he can't," insisted
the other. " But I tell you he did," said
Tommy. It is not what the police are supposed to do that Hodge sees, but what in
fact they actually do, which is not the same
thing.

There was a bazar fire near the house of a

friend of mine, which was put in danger. My
friend wanted a police inquiry as to who was
to blame. I expressed a doubt whether the
culprit would be detected. " Oh, I don't care
for that. Let them be a day or two in the
hands of the police, and it will be all right."

Power is a property that can readily be
transmuted into money, and no doubt it
often is.

I do not say that there are no police officers
who use their powers only as they are meant
to be used. But Hodge looks on the police
officer as an extortioner and a bribe-taker, to
be avoided as such. If by chance he meets
one that is honest, he is as much astonished
as I was in a certain cathedral which I was
being shown over, when I met a verger who
refused a tip.

The magistrate is the next personage. He
is the authority who revises the proceedings of
the police, and to whom complaints may be
made of the conduct of landlords, neighbours,

private persons, public bodies, and officials.
During my whole career in India I have held
the office of a magistrate, and have often tried
to imagine what things are like, and how they
go on, in my absence. In my presence I see
every one courting my favour, and eager to
carry out my wishes. Is this loyalty, or is it
hypocrisy?

In the district I now live in, with a million
and three-quarters of inhabitants and 5500
square miles of area, there are some eight
officials who may entertain complaints. We
hope in time to raise the number to ten or
twelve. The magistrate who has to receive
and sift the complaints of Hodge and others
is a man burdened with many public duties,
some of which involve money responsibilities;
often, too, with domestic and social cares and
worries. The climate, for all but about three
months in the year, is unfavourable to bodily
exertion, and in many places destructive of
health. It is no unusual thing in a sickly

season to find all the people of a place weak-
ened with sickness, and in this state having
to struggle through their heavy day's work.
The married men have sent away wife and
children, or, worse, see the wife and children
they cannot afford to send away pining before
their eyes. In small societies squabbles are
frequently going on, which are in a way useful, occupying the mind and keeping off ennui,
but which, in another way, wear out the spirit
and destroy its elasticity.

All this Hodge does not know. He looks
on the magistrate, European or native, as a
very great man with mighty powers. The
great man himself is in the centre of his
system, and, like a parson in his pulpit, has
always the last word. Arguments and evidence on both sides are laid before him, and
he finally chooses which he will believe. The
cause which he favours is right, and, unless
his decision is upset by superior authority,
which happens in only a very small percentage

K

of cases, is thenceforth supported by the whole power of the British nation, civil and military. The law as it exists, with its provisions, procedure, and precedents, binds his conduct, and the old methods of the witty Kadis which are related in the pleasant old Persian tales are not for him.

Hodge must take an interest in the magistrate, because, however much he may have resolved to have nothing to do with the courts, some neighbour or enemy may force him there; or, again, he may have himself been so injured that he is obliged to seek redress from the magistrate.

It is written in the law, in many sections, "If A has done so and so to Z, then A will be punished in such a manner." Hodge has, however, before lodging a complaint, to consider in the light of practical experience, "Will he be punished?" and, "Is the game worth the candle?"

The first thing he has to consider is the

time and trouble it will cost him to reach the magistrate. The magistrate's headquarters are five, ten, twenty, or even thirty miles off, and Hodge may find on getting there that he is out on tour and has to be followed a further distance. The journey has to be made over bad roads or perhaps across country. Hodge, unless he lives very near, may look forward to an absence from home of three days or more for every visit he makes to the magistrate's court. I remember, when I was at home on furlough, being in a small village about eight miles distant from the county town, with which it was connected by a good road. The parson was discussing a village rowdy who every now and then broke out in a drunken fit and created a disturbance in the street. The parson said, "I have more than once been inclined to prosecute him myself, but I should have to go in to Salisbury and give evidence, and who is to pay my expenses?"

I was much struck with this view of the matter from the outside, which confirms my own impression that distance and expense are strong deterrents to Hodge, when all he wants is redress for the particular injury he has suffered. Moreover, the time has been chosen, not by him but by the enemy; and if, as is usually the case, it is a critical time in field-work, such as ploughing-time, or weeding - time, or harvest, when every hand is busy and every hour precious, the chances are that Hodge will bear and do a good deal before he will give up the time needed for an expedition to court.

Then, if he is the only adult male in his house, he will have, when he goes, to leave his wife and children unguarded in their flimsy thatched hut with walls of mud or mat; the cattle and goats to be fed and watched; the unfenced fields with no one to keep off thieves or trespassers, whether four-footed or biped. If Hodge is strong

enough to take his change otherwise out of
the enemy, he will generally do so; if not,
why, he will still hesitate.

There are other things he has to think
of. What will the landlord say? Some
landlords object to their tenants' going to
court, as if they could not get justice at
home. What will the caste say? What
will the neighbours say? Will he be able
to get his witnesses to give evidence?
Hodge, we must remember, lives in a small
community, and there is a strong party
against him. Will his witnesses accept the
fatigue of the journey and the loss of time,
and will they brave the hostility of Hodge's
enemies for the sake of helping him in his
effort for justice?

Then comes the question of ways and
means. The Legislature has thoughtfully
cheapened the cost of coming to court, so
that a small initial fee is all that is necessary to start a case. But Hodge very quickly

finds that the court fee is the smallest part of the expense. There is first his own time, trouble, and food, and reasonable compensation for the time, trouble, and diet of the witnesses. If he does not satisfy them, he may whistle for their evidence.

Then the way to the court is beset by lawyers and lawyers' touts. There is a Solomon Pell always on the look-out for the Tony Wellers on their way there, not too proud to take pice from those who cannot pay rupees. The courts being few in numbers, each serving a large area, armies of prowlers are concentrated on their avenues, and Hodge has to run the gantlet through them. He is, especially when new to the business, bound to be taken in tow by some one who offers to guide him safely. Hodge in India cannot drop round to the office of his neighbour, Lawyer Smith or Jones, because India is a poor country, unable to support lawyers away from the chief centres of government, where the magistrates live. So Hodge falls into the hands of the first affable person he meets claiming professional knowledge, or, if he gets past the first, then the tenth or the fifteenth. This gentleman will find out what Hodge is prepared to spend, and arrange his bill accordingly. His first piece of advice generally is that it will never do to tell the truth, so that Hodge and his witnesses usually turn up in court with an ingenious but obviously artificial tale. Sometimes the simplicity of the witnesses betrays them, as when one turned to his principal on an unexpected question being put, saying, "You never told me what to say to that!"

However it comes about, Hodge in court is much more given to lying than Hodge among his neighbours, and I believe that is because he is handled before entering the court by unscrupulous tutors.

Well, we have brought Hodge within sight of the court, and may suppose that he has got his petition and power of attorney written, engaged a lawyer, bought the stamps and attached them, and is ready to enter the great man's presence. There is a good deal of waiting about, and at length he reaches the magistrate, and his turn comes. The magistrate is most likely an entire stranger to him, and he to the magistrate. They meet here for the first time. I know what the magistrate sees. What does Hodge see?

He sees before him a man seated at a table, pen in hand. Magistrates are of all sorts— European, Eurasian, Hindoo, Mahomedan; old and young; stout and thin; athletic and puny; healthy and sickly; patient and hot-tempered—in short, every variety of man. But one point they have in common—all are in a hurry. Hodge would like to squat on his hunkers as he would in his own village, and tell his tale in a roundabout and leisurely way, beginning somewhat as follows: " Well, ye see, sahib, it's like this here. I have a field

on the west side of the village. It's a field
my father had before me—been in our family,
oh—ever so long !" Most often, however, he
has learnt off a yarn which his legal adviser
has taught him. He appears before the
magistrate and begins firing off his story.
He is pulled up, made to stand at attention,
and hurried through a form of oath. If not
too much confused by this interruption, he
tries again, and is snapped up at once, and
brought to the point. His way would take
an hour, and the magistrate has only five
minutes for him. He often gets flustered,
and, forgetting the story he has been taught,
blurts out the truth. Very often, however,
he is questioned superficially, and the concocted story may pass. Occasionally he tells
the truth from the outset. Hodge going to
court is very like a schoolboy going up for
examination, and has a schoolboy's contempt
for his examiner's brains. But Nupkinses are
not common among the trained and ex-

perienced servants of the State, and the
practised eye of the magistrate searches deep
into the story brought before it by Hodge, if
only by a glance or two. Many a man is,
unknown to himself, protected from the
malice of his enemies by the good sense and
intelligence of the magistrate. When that
official is a very young man, some simple traps
are generally laid to draw him off his balance.
One is to say: "I said I would complain to
the magistrate, and he said, 'What do I care
for the magistrate? That for the magistrate!'" the intention being to cause the
magistrate to say, "Oh, he doesn't care for
the magistrate, don't he? We'll soon see about
that!"

Hodge has still many risks to run. It is
a favourite trick, when the injured person is
known to have gone to court, for the other
side to race him in, and even get the start of
him with a cross-case. As the wooer who is
not in love is often the most persuasive, so is

the petitioner who has not been injured often
the most plausible; and where the magistrate
knows nothing of either side, it is a toss up
which of the two gets the very important
advantage of the start. Then if the magistrate
be ill,—weak with fever, or out of sorts with
liver, or exhausted after a long journey or
with overwork; knocked up with the heat;
in a hurry to get on to his other duties, or
perchance to sport or private business, or is
of a fiery disposition,—Hodge's reception is
discouraging. I have been told of a magistrate who so furiously glared at the approach-
ing petitioner, and in so fierce a voice asked,
"Well, and what do *you* want?" that he
timidly said, "Oh, nothing, sir!" and
vanished.

If Hodge has succeeded in taking out a
summons and getting a day fixed, then comes
the process fee to be paid, and the tipping of
the clerks and court servants. It is illegal, of
course, but it is always done. Hodge calcu-

lates that on the whole it pays better to be on friendly terms with the court officials than to make enemies of them all, as he would by refusing to pay or complaining. A friend of mine who was stationed on the border of a native state once met a man who had migrated into it from British territory, and asked him how he was getting on. "Oh, much the same as before," he said. "But don't you find it harder to get justice over there?" he asked. "No," said the man, "I can't say it costs more." "But have you not to pay the *hakims?* On our side you did not do that." "Oh," says the man, "where I am now we pay the *hakims*, on your side we used to pay the clerks, and it comes to much the same thing."

Then comes the business before the trial, rallying and tutoring and keeping in good-humour the witnesses, who have now gained in value since they have been named, and doing what he can to prevent their being bribed or threatened off by the opposite side.

Then there is the journey to court, sometimes repeated many times, with or without witnesses, and finding out and meeting the case of the other side.

Altogether, long before the case has been decided Hodge has probably dipped deep into his little store, and gone dangerously close to his margin of credit. As only one side can win, perhaps he loses in the end; and if he wins, he may find that his enemy has lost, fine included, considerably less than himself —a Pyrrhic victory!

"Why does Hodge ever go to court at all?" one is inclined to ask. That we shall come to later on.

The civil court differs from the criminal court in two main points: one, that it is less serious, touching property rather than life and liberty; the other, that it is more dangerous, since a man may be condemned behind his back, which in the criminal court is never done.

No doubt care is taken to have notices served, but this opens up a wide field of profits to the process-serving establishment of the court. Hodge dreads the expense of the civil court, and its delays.

As for expense, Hodge has to engage a pleader, and to pay costs on costs, whether he is plaintiff or defendant, whether he wins or loses. The words of a Scottish poet apply very well:—

> "He took her three geese to get it begun,
> And he needit her cow to carry it on."

As for delays, his case may drag on for years. When a decree has been got, he is not at the end; for then begin a series of appeals, reviews, and revisions. It is the way all the world over. To quote again from the same author:—

> "In vain did the writer lad promise success—
> Speak of Interim Decrees, and final redress;
> In vain did he tell her that judgment was comin'—
> 'It's a judgment already this Soumin an' Roumin!'"

In India, for economy's sake, cases are tried in the first instance by low - paid judges ; and to prevent injustice, appeals and revisions are allowed very freely. But what interest has Hodge to lose that will compare with a cheap and swift settlement? Unless his land is concerned, Hodge seldom troubles the civil court, though he is often dragged there.

Then there are the process - servers. For economy's sake they have small pay, and, unlike the sheriff's deputies in England, they are not under any check of close supervision or of public opinion. A process-server has to serve a notice, perhaps twenty miles away from the court, on one who can neither read nor write. If he has been paid by the other side to report that it has been served but at the same time not to serve it, they will provide plenty of false witnesses to prove the service.

Lastly comes the execution of the decree—

the seizure of household goods, cattle, dishes, clothes,—even doors,—grain, and everything movable, and their forced sale at a loss; the seizure of one man's goods or land for another man's debts, which involves anxious and uncertain efforts to reclaim them; the submission to the court of false lists of things seized, articles being omitted—all illegal, but constantly done.

For Hodge, the civil court too often means ruin, and his feelings are much the same as those of Outram's " Grannie " :—

"The Priest spak' o' Job—said to suffer was human—
But she said, 'Job kent naething o' Soumin an' Roumin.'"

This is one of the joys of civilisation to which we have introduced Hodge, and the only thing he can say in its favour is that at least it keeps off other evils!

I shall never forget my experience when I was mobbed by crowds of anxious-eyed Hodges, who had been encouraged by an official to bring suits which in the end were given against them. "You, the *hakim*" (magistrate—not myself, but my predecessor in office), "told us to do it. We sold and pawned all we had—cattle, dishes, clothes— and borrowed money to buy court-fee stamps with. What have we got by it? We have lost the land we sued for, and everything else besides." It was a warning to me, and I am always most careful never to advise a man to go to law.

OTHER CLASSES

OTHER CLASSES.

HAVING finished what I have to say about Hodge, I will now shortly glance at the other classes of the community. There are first the great army of producers and distributors — the fishermen, boatmen, carters, milkmen, weavers, oil-pressers, makers of shoes and other garments, blacksmiths, potters, carpenters, goldsmiths, barbers, and other village servants, grocers, dealers in cloth and other wares.

In the great centres, where workshops and factories on an extensive scale have been set up, large bands of trained work- men and women have been organised. These have been more or less imbued with English notions, and strikes are not unknown.

I cannot say that I have any very vivid dreams of these as distinguished from Hodge. Many weavers, thrown out of work by Manchester, have taken to cultivation, while such men as skilled carpenters, for instance, are hardly to be got.

The chief characteristic of indigenous industry seems to be that it is all hands and no head—all labour and no capital—all instinct and little thought — power of imitation but little originality; and more patience than energy.

In every industry there is a capitalist who advances funds and materials, and who receives back the finished product. He generally holds the skilled artisan in his debt, doling out to him enough to keep him alive. This capitalist, however, is a trader, not helping the artisan with direction and advice, but merely exploiting him. Herein the artisan differs from the beast of burden, which has not to consider how to supply its own wants or those of any

one else. There are the same general characteristics, the same minute budgets; the same
problems of caste, society, domestic affairs, &c.
Although not so firmly tied to one spot as the
cultivating ryot in a closely settled district,
the artisan in the country is nearly as station-
ary. In the towns houses are to be had for
hire, and people can therefore freely change
their residence at their convenience; but in
the country a man has to build his own house,
and for this he needs a site. Over this site
much thought has to be taken, for it is not
a comfortable position to be a tenant at will,
liable to quit on notice at any time. He generally gets a lease, which secures him in the
land, and this anchors him. Nearly all arti-
sans and producers on an estate give to the
landlord of their skill and its fruits, whether
it be in the terms of their lease or not—the
fisherman of his catch, the weaver of his cloth,
the shoemaker of his shoes; the potter, the
milkman, and others of their products.

Passing from these, we now come to the two classes who work the springs of all industry in India—the money-lender and the landlord.

The money-lender is a man with a very bad name — indispensable, but hard to endure. When I first came to India I was, as all Britons are on coming to this ancient and conservative country, somewhat radical in my notions, by which I mean that I underrated the force of custom and of circumstances, and perhaps overrated my own capacity for managing other people's affairs.

I have slowly taken in the fact that the money-lender not only has his bad points—not only is too often an extortioner and so forth—but that he has also his uses, for the sake of which his bad qualities are borne with, and his grievances and difficulties which drive him to be what he is.

In fact, it is with the money-lender as with most other men, that he has a good deal of human nature in him. His acts and notions are to a great extent determined by circumstances, and he looks at things from a point of view of his own, which is different from that of his debtor, and from mine. With the money-lender, as with Hodge, the centre of interest is the family budget. He has so to regulate income and expenditure that the income exceeds the expense. The expenses, apart from business, do not differ materially from those of Hodge. It is the method of business and the source of income that are different.

In the first place, " Hunks," as we may call him (money-lender is such a long word to repeat often), works not with land but with cash. Having no land, he cannot so readily borrow, and must provide his own capital.

Hunks's capital is ordinarily limited, and the necessity that he and his family should live out of it forces Hunks, as it does the small man of business all over the world, to be very hard. If we bear in mind how small a capital many a family has to live out of, we shall, I think, be less ready to condemn its careworn head for his skinflint proceedings with his debtors, who are less interesting to him than his own family. His back is at the wall. He has no money-lender behind him to support him when his venture miscarries; and he must therefore screw out of his little business every farthing he can, suffer who may.

Hunks has his trials. When he has lent money to a debtor, he cannot recover principal or interest without an immense expenditure of time, labour, and perhaps money. He rides round on his pony to collect his little debt ten — twenty times, always to receive the same answer, "Not to-day : come another day." His debtor perhaps absconds, carrying off a morsel of the precious family capital, and leaving behind a bond not worth the paper it is written on.

Hunks has his ways of making little more.

He deals in running accounts, of kind and credit. Like the American editor, he believes rather in interest than in "principal."

Hunks lends and receives in kind as well as cash. I once saw an account which was worked out by the man who made it, in court, in which he tried to prove that for a loan of three bushels, made a year or two before, there was due to him a return of thirty. Then he buys with one measure and sells with another. I have heard of a trader who, when any one came to sell, used to call for the *burra bou* (elder wife), and when any one came to buy, for the *chota bou* (younger wife), thereby directing the people inside to bring out the large (*burra*) or small (*chota*) measure. Another used, in the same way, to call for Bechárám and Kenárám, which are proper names, and were used to denote selling (*becha*) and buying (*kena*). We introduced in one place cast-iron standard measures, which were largely

bought by the people, and discovered imita-
tions of these being secretly introduced that
held more, and sold for five times the price.
The selling measures are often lined with
mixtures to make them hold less. I have
been told of more lining being taken out of
one measure than it could hold when broken
up and loose.

Hunks makes his debtors sell to him at less
than market rates, and the interest on debts
that he charges is often 30 or 40 per cent a-
year, and sometimes double; besides that, the
principal is falsely stated, and little presents
of oil, milk, salt, and firewood outside the
account are taken when he pays one of his
visits.

For recovering his money Hunks has little
confidence in the law. A civil suit means
costs from beginning to end, and a very
doubtful chance of realising, even if he gets
his decree. He believes a good deal more in
the limbs of the law, and has more intimacy

with the subordinates of the police and of the civil courts than with their masters.

What Hunks needs for successful operations is a good screw. In some way he must have a man in his power. Sometimes this is done by getting him into debt on such extortionate terms of interest that the debtor can never hope to get free. Sometimes he gets hold of Hodge's land, and makes him work on as an under-tenant at a rack-rent. Sometimes the mere fear of not getting credit when it may be urgently needed is enough. Since the days of Joseph, those who have continue to drive hard bargains with those who have not and who are in extremity. Hunks has certainly this to be said for him, that he and his have succeeded in laying by some at least of that very scarce commodity in India, capital, and he may drive bargains because he holds a strong position.

When Hunks has once fixed his tap in Hodge, or in his fellow-producer, Hodge be-

comes henceforth the bee that makes honey,
not for himself; the ox that bears the yoke,
but not for himself. Hodge's compensation is
that in hard times he will have some one to
keep him in life and hold his head above
water.

Hunks is a person, and not, like most banks
in England, a corporation. In India there is
no law of primogeniture, so when the little
Hunkses grow up, marry, and have children,
the burden on the family capital increases.
When at last Hunks dies, and the family
breaks up, each member taking his share of
the capital and of the debtors, farewell to easy
circumstances, both for the younger generation, each of whom has to live out of a share
only of the old capital, and for the debtors,
who have to be screwed harder and harder to
provide the wherewithal.

All his life Hunks has to look forward to
this time in the future. Hodge may hope
that, when his sons are strong enough to

work, they will be able to expend their labour on more land. Hunks has no such prospect. No one is likely to give him more capital for his sons to manage, and make money out of. What he gains, that they may look for, and nothing more. Therefore there is little hope that, while things are what they are, Hunks will cease to deserve his reputation as a skinflint and an extortioner. He cannot help himself. Needs must when the devil drives. And this will continue to go on until money-lending transactions are conducted out of savings instead of from the living fund.

We come now to the landlord, who shares with Hunks the office of Providence to Hodge.

A person of the well-to-do landlord class is generally uneducated, having been a spoilt child from his youth up. In India education and accomplishments have been hitherto regarded rather as a means of living for the poor than as a necessary ornament for the wealthy. Like the old king who was "above grammar," they are too great men to need education.

The name of the landlord in most large estates is legion. With the exception of a comparatively few estates, which custom forbids to break up, the property is generally, by the ordinary law of inheritance, split up into many shares, the owners of which are of many varieties, old and young, men and women, rich and poor. Each owner has, and many of them exercise, the power of selling, subletting, or mortgaging his or her own interest. The person so acquiring a share in the estate is often a speculating stranger who does what he can to increase his profits at the expense of his partners. Rights are sublet, and again sublet, so that the records of titles in a large estate often present a most bewildering number and variety.

The custom of the country is that, while rights are many, the management vests in one leading partner; and there are heads of households the members of which number hundreds, and whose affairs are nearly as important as those of some small States.

What with keeping the peace; preventing the break up of the estate; guarding its interests as a whole in relation to tenants and neighbours; watching the doings of the various partners, their assigns and lessees, and the servants of the estate; arranging for payment of dues and wages, and for the distribution of income to the shareholders; providing for the maintenance of poor relations and widows, for marriages and other family ceremonies, and for other affairs too numerous to relate, the manager of the family has enough of politics within it to take up his whole attention.

There is a saying, that one generation makes a family, the second keeps it going, and the third breaks it up. It seems to fall to pieces by its own weight. Even where there

is a capable manager, who is trusted and
obeyed by all the shareholders, sooner or
later comes the dreaded time when, either
owing to the death of the manager, or other
causes brought about by time, dissension
raises its head, and there is a demand for
separation.

I remember an old family which was kept
united for a very long time by the dread of
a curse laid by the founder on the first member who should break its unity. At length
a scapegoat was found in a widow, who had
a tiny share for life, and was induced to
sue for its separation. The opportunity was
seized by all the principal shareholders, who
deluged the civil and criminal courts with
litigation, which, for all I know, is going on
yet.

Family dissensions in great houses are a
national calamity, and have been before now
deemed of sufficient importance to justify the
interference of high officials. The last year

of Sir Ashley Eden's term of office as Lieu-
tenant-Governor of Bengal was distinguished
by successful efforts to bring peace into leading families; and to make such efforts in a
smaller way is part of the duty of all Government officials.

The head of a great family, like every
other head of a family, has before him the
problem of the budget as the most persistent
and troublesome of all his problems. Most
landlords, being small men and not big, have
it in an acute form. Few landlords are
wealthy. If the income is large, so also are
the outgoings. Most families inherit debt;
and what with this and litigation, and the
payment of charges on the estate for revenue,
management, maintenance, and the like, there
is more often a deficit than a surplus at the
end of the year.

As the family grows, the problem of house-
room grows too. The wealthy members generally live off, and build separate houses for

themselves, leaving the old family residence to go to ruin, a rookery for the poorer members who have nowhere else to go. The only occasions when the others visit it is to worship at the old family shrine, when there is often a struggle for precedence, accompanied by fighting.

The maintenance of the poorer members is a charge on the head of the family, and evidently presses on him severely. I see men of the highest position striving with all their might to get poor relations provided for in the service of Government as clerks, constables, or messengers even, if nothing better is to be had — anything to lighten their shoulders of a part of their heavy load.

The manager of the family generally acts through others, and is not so much himself a man of original force as a centre of influences, chiefly of relatives—mostly females—and servants. Ladies have thus been known successfully to manage great estates. The pressure is brought to bear on the manager, and he acts. The great landlord is often disappointingly careless and incompetent when lifted out of his natural position, and placed, say, on the bench, where he is expected to act and think for himself. You will see a smart and penniless lawyer who easily outshines him in eloquence, argument, and quickness. But it would be a great mistake to estimate his importance by the figure he makes on such occasions. When it comes to a question of influence, the apparently lethargic head of a family will be listened to and obeyed, while the brilliant orator may rant in vain. He speaks with a wisdom and a weight far more than his own for the great constituency he represents, while the other is comparatively an adventurer.

As to the towns, I have had something to do with the people in some of them, though not so much as with Hodge. They are, as being a small fraction of the population, of

comparatively little importance; but yet in
some ways they bulk largely before the world.
For one thing, they are formed into larger
communities than the rural population—communities that have rubbed shoulders so much
with the world and with one another that
they are more intelligent and quicker witted
than the villagers. Among them are arising
professional and business men of all the call-
ings which make and are encouraged by capital
—men who can write and talk and organise.

Then, again, there is generally a town
at each centre of Government. The townspeople are thus better known to Government
officials, and know them and their ways
better. They see the Government officer
going out for his daily ride, taking his
amusements, and attending office. No doubt,
if inclined that way, they can gossip with
his servants, tradesmen, and clerks about his
affairs.

To the Government official the people of

the town in which he lives are apt to have the importance which the near always has when compared with the distant. He is tempted to judge others by their life, and their affairs occupy a disproportionate share of his time. Since their manners and customs are modified by the courts, lawyers, and police in their midst, any one taking them for a sample of the country would have a wrong impression of it. I have often thought how typical of the town-moulded mind is the Calcutta newspaper, which, in a country whose law courts are teeming with tragedy of the most desperate kind, passes most of the tragedy by unnoticed, but finds a place for a report of any petty assault case or accident that happens in the town itself. The local incident has come under the notice of the reporters, and the others have not. In this respect the town community is little in advance of the village community, its members being ignorant of all beyond their own narrow circle, and unaware what a small proportion of the whole country that circle is.

To a certain extent this state of mind is paralleled in Europe, where both the London cockney and the Paris bourgeois are eminently local, but there is a difference. One difference is that the townspeople of Bengal live as it were in a glass case, and have no occasion to consider the great national problems which take up much of the attention of every citizen of England and France. I remember receiving a deputation from a certain town, which suggested that a great deal of money might be saved by abolishing the police, who were no longer needed. I asked why they thought that, and they said because there were no burglaries or thefts now. I said they might as well break down their garden fences because the cattle did not get in. The Bengal town is a well-walled garden, and the people do not even know that there is a wall, and what it is useful for.

Again, unlike the English towns, those of Bengal are full of improvements and institutions which are so far from local, that most of them were forced on the towns against their will. How their secret inclinations tend —what they would do if outward influence were removed — is a matter of doubt. My own impression is, that a good many of those institutions that are now apparently their pride would collapse like the *soufflé* pudding which "got up and sate down."

I do not think the towns will advance far, unless the villages advance with them. They have been pushed far ahead of their natural place by the outside influence of Europeans.

These are my dreams of the classes of Indian people other than Hodge.

ARISTOCRACY

ARISTOCRACY.

A CLASS about whom I have often had my dreams is the aristocracy—those who call themselves the "gentry" or respectable classes.

In Bengal the gentry are determined as a class by birth, and not by wealth or ability. The poor man of high caste, even though in a low position, despises the rich man of low birth, who is not in his eyes fit to eat with, or to intermarry with. Pride of birth is common to all parts of the world. We Britons have it strongly developed among ourselves, and the claims of those who have birth are, in the main, acknowledged by those who have it not, both in Britain and in India. How do our

Indian men of birth take themselves? What do they think of their duties and privileges? What do they desire to make of themselves?

I believe that our British notion of an aristocracy substantially agrees with the meaning of the old Greek word—the rule of the best. The controversy over the House of Peers is generally centred round the question of its merits. Those who denounce endeavour to prove that its members are useless, and those who defend point to what they have done and are doing for the nation.

The man who receives honour and comes to the front, is he who adopts the motto of the heir to the Crown—" I serve." The chief subject is the " Prime Minister," or head servant of the nation. Service, though often well rewarded, is not always given for the sake of reward. The nation has many wants, and many servants who are working to supply those wants, of whom it is not aware, and of whom a very large proportion get no reward of wealth or honour while they live, or even afterwards.

Those who receive honours as the "best" in Britain are those who are believed to struggle and strive for her service. They labour in many fields—in battle, on land, on the sea, in the mine, in the tropics, at the poles, in the backwoods, in the study, the hospitals, the law courts, the senate, the school, the pulpit, the studio, the workshop and factory. The British nation is rich and powerful because she is well served. No matter how great the risk, or severe the service, let her call for men, and they leap forth, ready to do or die. When she finds a man she honours him, as the Seigneur did Dunois :—

> "Puisque tu fais ma gloire,
> Je ferai ton bonheur."

In these days, when there is a party of Indian politicians desirous of replacing with natives of the country the small band of British officials who still take part in the gov-

crnment of the country, curiosity must be felt
about the minds of the class from among which
these substitutes would be taken. I am not
without experience of that class, having had
thousands of them under me as subordinates,
done business with thousands, and investi-
gated the disputes of thousands more.

Now that they have thrown their cap on
the stage, and have become competitors for a
prize which, if they win it, will make them
guardians of vast interests, politeness and
reserve must give way to sincerity and can-
dour. The men of this class must be regarded
with a very different eye, and scrutinised with
greater severity than would have been necessary, or even right, when they were nothing
more than bystanders or underlings.

To begin with, Who are, in the eyes of the
active politicians, the "people of Bengal"?
We hear sonorous periods about the vast population—the countless millions—who are crying
for "justice." It appears to me that this is

rhetoric to catch the ears of the British nation. The politician of Bengal is no more thinking of sharing with Hodge and the masses of the population the fruits of his agitation than the Australian squatter thinks of taking his sheep, cattle, and horses into partnership.

In England, the "best" spring from all grades, classes, and races; in Bengal the higher castes—a small minority of the population— claim that capacity for and the right to rule are to be found within their limits only among the people of Bengal. This is their theory, whether it be correct or not. Their dream is a caste—not a national—dream.

This dream of theirs—what is it? How will they work out their destiny if given their own way?

We have in England an expression describing the men by whom her prosperity has been built up, as "captains of industry." These men have forced themselves and their dreams to the front, most often in the teeth of resist-

N

ance from the Government and the public. By their moral and physical, no less than their intellectual qualities, they made their opportunities, and utilised them.

In India many great institutions and industries have been built up,—the Government, with all its vast machinery for the service of the public; the railways; the great industries of tea and indigo; the factories and mines. But it was Europeans who made them, and carry them on. The natives of the country have no doubt supplied most of the rank and file, and a share of the national means, but very few of the "captains" have come from their ranks. Even among nations of Asia, the comparatively small sections of Parsees, Marwaris, Jews, and Armenians have produced much more business enterprise than the overfull ranks of the Bramins and Kayests of Bengal.

It would be going too far to say that because the high castes of Bengal have in the

past produced few "captains of industry,"
therefore they cannot be expected to produce
many in future. But the qualities which go
to make a "captain of industry" have a way
of forcing themselves into notice. If people
say, as some do, that the natives have no field,
why is that? Men with the right qualities
will make their own field, as sure as water
finds its own level.

The dream that governs the British race is
one of energy and of conquest — the strong
man rejoicing in his strength, and employing
it for the conquest of men, of wealth, of disease, and of the forces of nature. The dream
of the native of Bengal is rest, freedom from
want and from worry. He does not exert
himself for the love of it, but only to save
himself from want or trouble. To earn a living; to have no fear of an attack by his
neighbours; to keep on good terms with the
authorities and all who can do him harm,—
these are the motives that in India drive a

man to work and exertion. He does not "re-
joice in his strength," but works because he
must, and ceases as soon as he dares.

It is here that the British and the Indian
dreams are at cross purposes. We dream—the
great mass of us do, though the cynics and
superior persons pity the delusion — of vast
springs of energy bubbling up against the
pressure of adverse circumstances, and ready
when these are removed to gush forth. The
Bengali rhetorician, catering for English au-
diences, conjures up lively pictures of the
same state of things in India, the adverse
circumstances there being a handful of British officials. Remove these, and the fountains
will spout forth. Among themselves, however,
this is not their dream, and I doubt if they
have any but the dimmest notions of what is
to follow. We believe that we owe the great
place British institutions and British men
have attained in India not so much to clever-
ness and calculation as to moral and physical

energy. In intellect we find rivals among the people of the land, but what we are rich in is motive power. I do not think they realise this fact, and I believe that, not realising it, they, without the command of motive power, seek to be intrusted with tasks which cannot be performed unless with its aid. A Scotch friend of mine, managing a jute-mill, said to a native who was arguing about the Ilbert Bill : " Do ye see this mill, Baboo ? How long do ye think ye could run it without a break - down ? I'd give ye two days ! " This is a very common dream among us Anglo-Indians. Our British aristocracy were always noteworthy for their energy. In warlike days it exhausted itself in fighting ; and in our more peaceful days it has been turned into, and utilised in, other channels. Other classes have entered into friendly and more and more equal rivalry with it, until the national energy has overflowed into every continent. The highly paid

posts — the various emoluments that have
followed British enterprise all the world over
—are accidents; the work was the end, these
are but necessary means to the end. I fear
the dreams of the native classes are fastened
not on the work but on the emoluments.
Our end is their means; our means their
end. They do not understand our dream,
and they dream that we are also dreaming,
like themselves, not of the work, but of the
emoluments.

To conduct the business successfully, we
must see, first, that it is well protected;
second, that there is sufficient motive-power;
third, that the machinery is kept in good
working order; and fourth, that economical
and sufficient results are got.

I fear our "no foreigner" school of politicians have very little notion what a rough
place the world is, and how troublesome a
team they would have to drive, even within
the province, supposing that no one from

outside jumped on the box and snatched the reins from their hands by force. I fear, too, that their end being a different one from ours,—the emoluments being the end, and the work a mere means of obtaining them, —the business of the country would not be done thoroughly and successfully, and might fall into confusion.

The politicians, no doubt, reply to this that it is a mistaken idea, and that there are plenty of men endowed with honesty, energy, zeal, and every quality needed for the successful conduct of the country's business. I do not think any practical man has proposed to give up British protection. The idea of most of them is that Britain should mount guard while a fair trial is given to the plan of the politicians — a division of labour suspiciously like that proposed by John Leech's little girl to her baby-brother by the sugar-barrel, " Now, Johnny, you drive away the naughty wasps, while I eat the sugar."

I doubt if the politicians have any definite plans in their head as to how the "trial" or experiment is to be carried out, if it is ever allowed to be tried; whether they have ever thought of any further change in a machine designed and intended to be made of one material than the substitution in some vital parts of a different material for which it was not designed; or whether any arrangements will be made for a break-down gang in case this huge experiment on the "vile corpus" of an Indian province, its people and their interests, should not answer the expectations of its promoters. I have had many dreams about this experiment which some people want to see made, and the following is an effort to present some of the most lively of them in a short space,

THE BENGAL REPUBLIC

A DREAM OF HOME RULE

THE BENGAL REPUBLIC.

A DREAM OF HOME RULE.

IT is the birthday of a nation. The President of the new Republic has just been elected, and faces for the first time the Legislative Assembly with which he is to share the honour and responsibility of ruling a mighty people of 70 million souls. A Ministry has been formed. The ministers, officers, and members of the Assembly are in their places, and turn silently, on this historic occasion, to hear the first words that will fall from the lips of the new chief of the State.

As one glances along the rows of members,

it is apparent, from their intellectual fore-
heads, pale complexions, and delicate hands,
that they are more familiar with the study
than with the field.

The President, in opening the session,
begins by saying that freedom has come
upon Bengal suddenly, unexpectedly—almost
before she was ready for it. Not without
misgiving — even remonstrance — have the
leaders of the nation witnessed the with-
drawal from their country of those British
troops on whom they had hitherto relied for
the maintenance of order. But, since the
British Government have insisted that those
who make and carry out the laws should also
be responsible for keeping order, the leaders
of the nation have accepted responsibility for
the maintenance of order, stipulating only
that, on payment by Bengal of the cost, the
frontiers shall for some time longer be
guarded from invasion by British troops.

Thus they have at length come forth from

under the protection of that great nation
whose fostering care has advanced them to
so high a place among the peoples, while
still retaining her friendship and good
wishes.

An army has been formed of good material, the men being recruited from among
the warlike races of Upper India, and the
officers being mainly obtained from among
the flower of their own youth, unrivalled in
intellect, and well trained in the athletic
exercises of Europe—*mens sana in corpore
sano!*

Hope, as is right on such an occasion,
prevails; and as minister after minister rises,
and in flowing periods gives an account of
the department over which he presides, one
draws a contrast, not wholly to the advantage of our own people, between these finished and ready orators and the European
rulers of old, who were, nearly all, halting of
speech, awkward of expression, and slow of

utterance. A full treasury, a well-balanced budget, a well-drilled army, and departments in a state of high efficiency, are sufficient to account for the satisfied buzz and the happy smiles of congratulation which follow one another over that sea of faces, like the waves over a field of standing corn swept by the wind.

The clouds fall, and open on a different scene.

We are in a meeting, it may be of a district board or a municipal committee. What is this noise of wrangling that we hear? The foreigner has been got rid of, and yet we hear cries of tyrants and traitors,—strong and earnest denunciations, on the one hand, of oppression and jobbery; on the other, of conspiracy and sedition.

We listen attentively, and from among the babel of voices we begin to gather something coherent. There are two factions among the members at the meeting, and each side has a chief. The chiefs are grandees, much too dignified to sit in such an assembly themselves; so they have caused their followers to be elected members of the board, while they sit outside and pull the strings.

One disadvantage of what has been called " oratorical pressure " is the tendency of several speakers to hold forth at a time, and a consequent difficulty in hearing what is said. There is also much that is irrelevant in the flood, or rather floods, of eloquence rushing about the bewildered ears of the listener. Steady reiteration is, however, found to be given to the assertion, on the one side, that the chief of the majority is giving away appointments and contracts from corrupt and party motives; and that the officials and contractors favoured by him sacrifice the public good to the interest of themselves, their party, and chief. On the other side, it is urged as constantly that what the minority charge the majority chief

with doing is exactly what they want to
have a chance of doing themselves—that, and
nothing else, being the secret of their violent
language.

Up and down the country this violence of
language, these constant charges of jobbery
and peculation, are bandied to and fro. The
highest character, the most exalted position,
is not spared. No doubt grave accusations
and foul insinuations are the more lightly
employed that so many of the speakers and
listeners are lawyers, who import into lay
arenas the licence of the Bar. But people
outside the discussions begin to wonder if
there is not, after all, some fire with all this
smoke.

Again the cloud falls, and this time, as it
rises, we see before us a social gathering of
Scotsmen in Behar. It is St Andrew's Day,
when, tenacious of old customs, the sons of
Bonnie Scotland meet for one night to forget their exile. The feasters are few, and

they have not the prosperous appearance that used to be observed at such gatherings. Gloom and discontent are on every brow, and a dangerous look is in many eyes.

For these are the remnant of the once great community of planters and merchants who in former days claimed to be the leaders of enterprise in this part of India. The period of Native Rule which preceded the Republic witnessed the departure of many of their chief men, and the outflow has been still more rapid since; for under the Republic it has not gone well with them. Those we see here are nearly all men who would fain go and cannot, with perhaps a sprinkling of tenacious and brave souls who look on the present order of things as temporary, soon to pass away.

The broad Doric which prevails betrays the more provincial character of the greater part of those present. There is not wanting, however, a dash of the more refined but not less

genuine "classes." It is one of these latter
who occupies the chair. Careworn and sad,
little acquainted for many a day with prosperity, he is yet honoured as a man of ancient
family—one of the chiefs of the race—of
unblemished honour, high courage, and un-
wavering dignity. In the old times he was
well known as being always in the heart of
every movement for the welfare of the natives,
and even now he cannot join in the fierce
looks and bitter words of many a one among
his neighbours when the native is talked of.
From his eye there is ever a kindly gleam,
and mild words alone pass his lips.

Those who look at his grey hairs and worn
face, and think of what he has had to suffer,
feel shame to talk fiercely and angrily, when
he, who has a better right to do so, can keep
his mind free from resentment.

He has just risen to propose a toast. He
says that since the last anniversary of their
national feast - day an event has happened

which renders it necessary to change the form of this toast. Formerly it was "The Viceroy and the land we live in," but now for the last ten months we have been living under the rule of the Bengal Republic. We have been torn from our beloved country and turned into foreigners; and Bengal has got Home Rule. Since it is no longer permitted to drink the health of a British governor, and the sense of the managing committee is against substituting the President, it has been decided to alter the toast by omitting the Viceroy's name, and to make it simply " The land we live in." He then goes on to say that Scotsmen have never been wanting in a kindly and sympathetic feeling for their native neighbours. As things are at present, we have little reason to love the Government. We must never forget, however, that the Government is one thing and the People another. We are as much bound up as ever with the prosperity of the land we live in—nay (and here with a sad smile he casts his eyes around the thinned ranks and worn faces of his audience), all those who could leave have already left, and we who are here must sink or swim with the ship. He goes on for some little time longer in the same strain, and ends by proposing " The land we live in," which is drunk more heartily than at one time seemed likely.

The conversation is resumed, and, as is natural after a more or less political speech, it turns on politics. Partly because it comes natural to many, partly as it is not, like ordinary English, understood by the servants, the Scotch dialect is used more than it would have been in ordinary life.

" Ay," says one big-headed, sandy-haired man, " things is a' tapsalteerie e'en noo. Them that sid hae been keepit butt hae won ben."

" Deed, Jock, ye're richt there. Wha wad hae thocht ten years syne 'at you an' me wad be gettin' kickit aboot by a wheen 'braw, braw clerks in a ofish'!"

"Clerks, quo' he!" says a third. "They're far aboon the like o' that noo! Set them up! They're like oor ain hotch-potch, that has 'peas intil't, an' beans intil't, an' carrots an' neeps an' greens intil't!' There's M.P.'s intil't, an' judges intil't, an' managers an' secretaries an' a' intil't."

"Ay, an' there's generals intil't an' cornels intil't," says another.

"Hoot awa', man!" says the second speaker. "They're naebut like the curry they used tae gie us on board ship,—ae day it wad be ca'd Portugee; anither day it was Bombay; syne it wad be Madras,—ilka day a new name, but it was a' ae curry! Ye may ticket them M.P.'s an' cornels an' siclike names, an' dress them in braw uniforms, but, when a's said and dune, they're naebut a wheen clerks!"

"A'm thinkin'," says good-natured Jock,

who fears the conversation is growing too
fiery, "we wad be nane the waur o' a
sang."

"Ay, man," says another, "but I heard a
fine sang the ither day at Davie Anderson's,
an' here's the lad that baith made it an' sang
it." With this he pats on the back a blushing
youth.

"Come awa', Willie lad, an' let's hear yer
new sang. What's it aboot? I'll wad this
bap it's aboot yer sweetheart."

"Na," says the introducer, "there's naebut
ae thing we can think o' noo—politics! Man,
it's a fine sang, an' garred a' oor heads birl.
My haun' fair dinnled wi' rappin' on the table.
An' it's tae a graund tune tae—'The wee, wee
German Lairdie.'"

"What's the name o' the sang?"

"'The sly bit wee Bengali.'"

"Od," says one, "it'll be a rare ane! I'se
warrant he's sortit them weel! Let's hear't,
man!"

"Stop a bit," says the chairman. "I am only afraid it may rouse bitter feelings against our rulers. I gather from the title and the tune that the song is an attack on the natives of Bengal."

"It's a' that, sir; but ye maun jist ca' tae mind yer ain words in yer speech that the Government is no' the same thing as the people. We've aye been friendly tae the people, an' never said a word again' the Baboos till they cockit theirsels up in places whaur they had nae business tae be. We was frien's eneuch wi' them or ere they grippit what wasna their ocht. As lang as we were subjects o' ae Croon under ae Parliament, an' that oor ain, we wadna hae said a word. But we hae been sel't or gien awa tae them, an' they bood tae be our maisters,—an unco word, I trow!"

"D'ye mind, sir," says another, "what the Ulster men said when there was word o' Hame Rule for Ireland? 'Ye can govern us yersels,' quo' they, 'but ye hae nae richt tae gie us owcr tae onybody else tae govern.' "

" I can't deny," says the chairman, " that we have been cut adrift by the Government we owed our loyalty to; and I myself have suffered severely because I would not transfer it to our new rulers. Well, my lad, let's hear your song. It would hardly be fair to judge it unheard, at all events."

So Wullie Macneill rises and sings his song—

THE SLY BIT WEE BENGALI.

" Wha the deil hae we gotten for a king
    But a sly bit wee Bengali!
We learnt him first a sang tae sing,
    An' wow! but he sang sma'ly.
He was soopin stairs an' dichtin' shoon;
Whan we suppit oor brose, he'd lick the spune,
An' there wasna a thing but he'd hae dune,
    This sly bit wee Bengali.

Noo he's clappit doon in oor justice-chair,
    This sly bit wee Bengali;
An' his sib an' cronies gather there,
    An' hech! they blether brawly!

He has filled the land wi' writer loons;
Mortgaged the country an' bocht the toons,
An' on rich and puir he sets his thooms,
    This sly bit wee Bengali!

Oor chiefs lang syne were men o' micht,
    Wide waved their banners glorious;
Their bluid ran free for their country's richt;
    They fell, or stood victorious.
But I doot for fechtin' ye've nae great will;
Instead o' a sword ye'll wave yer quill,
An' for yer banner a lawyer's bill,
    Ye sly bit wee Bengali!

Oh man, but, Sandy, Jock, and Pat,
    Ye've a' been unco pally,
Tae loot the knee tae a cuif like that—
    That sly bit wee Bengali!
Then tak' yer laurels in yer han',
They're no' the fashion, tho' ance sae gran',
For wha the deil noo rules the lan'
    But a sly bit wee Bengali!"

As the song proceeds the audience listen with
that intent look peculiar to Scotsmen, and
the glare of enthusiasm shines in every eye.
Their bodies sway; their toes tap the floor;
they clink their glasses, drum on the table to

the time, and join in the refrain. General applause rewards the blushing singer, and all rise to drink his health.

The chairman, fearing least the passions roused by the song should become too violent, attempts to restrain the excitement; but in vain—it is too strong.

"Od, sir," says the second speaker, "oor forebears ne'er tholed what we hae tae pit up wi'; an', my word, neither will we!"

"Gin ye dinna like the new sang," cries another, the light of battle in his eye, "I'll e'en gie ye an auld ane!" and in a stentorian voice he strikes up "Scots wha hae!" in which all present join with a roar. When this is done, "Scotland yet" is sung with equal enthusiasm, the chairman at last letting himself go, and joining in. Most inconvenient things are these Scotch songs when the Scotsman is undermost in any part of the world. They make him restless and unfit for his place. As the toddy-tumblers are being replenished, and the buzz of talk goes round more loud and general than before, the vision passes.

Again the cloud lifts, this time upon a Cabinet Council. It is attended by all the chief ministers, heads of the great departments. They seem to be in trouble. It appears that the Budget forecasts are not being fulfilled, and the revenue is not coming in as had been expected. From one to another the Prime Minister looks, but in no direction does he find comfort. He hears of arrears of land revenue, and a diminished yield of excise, stamps, and customs; of a falling income and growing expenditure on the railways. He hears that trade is deserting the port of Calcutta, that some of the mills have been closed, and all the rest are running short time; that tea and indigo are ceasing to come down, and that the supply of jute is diminishing; that the coal-mines are closing, and industry in all its branches languishes. Why is this? Who knows? Every one has been diligent in issuing orders and reminders, and in calling for explanations.

The Home Secretary demands that the police be strengthened, since bad trade has caused much distress, and the criminal classes have been recruiting their ranks with men that it has reduced to ruin and desperation. The War Minister reports that the soldiers are out of hand, and openly show disrespect for their Bengali officers. They can only be kept steady by severe punishment of insubordination, and by the grant of fresh privileges and advantages. The Finance Minister says the long and the short of it is this : outgoings are rising rather than falling, and income is dwindling. This cannot go on. If we are not to land in bankruptcy, something must be done.

What is to be done is not so easy to say. The members of the Cabinet have studied political economy long enough to put aside as useless many expedients that would look

hopeful to less enlightened men. They can
but issue fresh instructions to all local officials
that every effort is to be made to get in the
revenue, and that expenditure is to be re-
stricted as far as possible. The cloud falls on
a group of very anxious men, who know they
are in a mess, and see no way out of it.

We flit with the speed of thought to a distant country place, where a group of local
officials are discussing the recent orders of
the central Government.

"Things are certainly beginning to look
bad," says the chief. "We have sent away
all the money we can spare, and are making
every excuse to avoid paying any out. Yet
the treasury is empty, and, what is more, is
not likely to fill. Something must be done.
There is not enough to pay us our next month's
salary. Oh yes! You may give reasons; but
will reasons fill the treasury? The fact is,
neither the outside public nor our own subordinates are afraid of us as they used to fear

the Europeans. It is very hard; for even when the Europeans were here we did most of the work, and I am sure we knew far more about the rules than they ever did. But there it is! When they gave an order, the thing was done, while we only get excuses and explanations. Come, gentlemen, can you suggest anything?"

Then says that shrewd man, the collector of income-tax: "In my department there are a good many in arrears. It would never do to press Baykanta Baboo, our patron. But there are Mr Wilson, and Moulvie Wahedulla, and a few others, on whom the screw might be put, and even some penalties might be got out of them."

Says the excise officer: "There are seventeen shops that have not paid up. Eight of them Baykanta Baboo's men have, and they must be let alone. From the others I will squeeze what I can."

Says the district police officer: "My men

are getting discontented. With all these idle and starving people about, there has been a regular outbreak of gang robberies and burglaries, and they are being worn out with extra duty. The men have had no pay for two months, and are pressing for it. They say they cannot live on air."

" When did they ever live on air ? " says the chief. " Let them live on the people, as they always did."

" Ah, but things are different now. The red turban does not inspire the awe it did, and my men have been several times refused food and lodging, and even beaten when they tried to take by force what they needed."

" Then start a good prosecution against some fat trader. You could keep a score of men for months on the proceeds."

" Fat traders are not so easy to find nowadays. It is bad trade that has brought us to this, and bad trade means lean traders."

"Anyhow, it is clear your men can't get their pay yet. They must wait."

" If they must, they must. Only it is my duty to warn you that you may expect a good deal of desertion. They say the robbers are better off than they are, and at least work for themselves."

" I can't help it. We must trust to the army. Now mind, gentlemen, if you catch a criminal—fines, not imprisonment; and if you get a well-to-do man, give it him hot, and try if you can to catch him in contempt of court. Then——" And here the cloud falls, and the vision flits to a country town, where we see a group of traders talking.

" What am I to do?" says one. " I am losing all the trade I had with Calcutta. The roads are all broken up, and freight of carts has risen. It is very difficult to get those stupid ryots to take lower prices, and unless they will, I cannot export produce at a profit."

" But why are the roads bad?" says

another. " I know the road-rates are high enough."

" Ah ! " says a third, " don't you know who is the great man with the board, and whose friends get the contracts ? Little chance is there of any complaints against them being listened to ! "

" That is not all," says another. " Formerly, when the station - master and goods clerk charged us heavy fees for letting us have waggons, there was at least an inquiry if we complained ; now it is no use complaining, for they are all some politician's pets."

" By the by, what was that accident on the line the other day ? "

" Rotten sleepers—more jobbery ! "

" I hear the railway are going to raise freights."

" They've done it already, crushing out what little business remains."

" Is it true that the Calcutta merchants

have lowered their rates for up - country produce ? "

" Yes. The politicians have been reforming the Port Trust and the pilot service, and big vessels won't venture up the Hooghly. Goods have to go down in lighters, and so they have had to reduce our prices."

" I hear you had a burglary in your house the other day."

" Yes, and one of my carts was looted too. When the police came, they were so lean and ragged and greedy that I thought it was the robbers back again. They can't be getting any pay."

" Are you bothered about income-tax ? "

" I should think so ! Are you ? "

" Oh no ; I'm on Baykanta Baboo's side."

" Worse luck, I'm not. I suppose it would be too late now to——" And we are whisked away to a room in a great house, where is being held a meeting of gentlemen — Europeans, Mahomedans, and one or two Hindoos.

As we enter, a European is speaking, and we catch his concluding sentences.

"So," says he, "it is understood that we are all in the same boat. Whichever one of the more numerous factions prevails, we are bound to be below. In the hard times that are coming we shall first be squeezed, and they will squeeze the very life out of us."

"What can we do?" says a native gentleman. "Who can resist the Sarkár?"

"Hark ye!" says another European, "sit close. There are different kinds of Governments, and I am much mistaken if this one is of the same kind as the one before it."

"Where is the difference? There is the law, the courts, the magistrates and police, the jails, the taxes, and the army—as before."

"The Government depends for support on four things—the goodwill of its subjects, the revenue, the police, and the army. Has it got the goodwill of the people?"

"No, I think not. Of course we ourselves

dislike it, but I doubt if any but the thick-
and-thin partisans of the ruling factions care
for it. They are getting offices and contracts,
and the public interests are suffering. Trade
is bad; many are in want, and want means
discontent."

"Then what about the revenue?"

"I believe the Treasury is very hard up.
Taxes are not paid."

"Good! What about the police?"

"They are certainly not in good order.
They are in arrears of pay and overworked.
Many are sick, many retiring or deserting.
Recruits cannot be got."

"Then there is only the army?"

"The army is untouched. It is paid
regularly, and is up to its full strength."

"Has any one seen it fight?"

"No, it has not been in action yet; but
its men are good stuff."

"What about its officers?"

"Very scientific, and well up in their drill."

"Would they stand or run away if they came under fire?"

"Of course one can't tell, but they are Bengali Baboos, and old instinct might be too strong for them. They would probably run."

"Then a fig for the army! We have not so much against us, after all. Don't let us wait to be eaten up piecemeal. Let us hold together and keep our own."

"I agree that we should resist," says the native gentleman, "if we have any chance, for we are bound to be ruined unless we do."

"Well," says the first speaker, "see here ——" And while they are consulting, they fade from view. After this, vision succeeds vision rapidly, and they are hurried and confused.

I see a great crowd — a mixed multitude of Europeans and natives — bound together by the common tie of fear. The little knot

we left discussing made up their minds to
open resistance, and are astonished at the
number of malcontents who have come to
join them. A leading statesman has come
down to prove how foolish and hopeless it
is to resist. He delivers an oration in his
best and most eloquent style. They say
quietly, when he has finished, that they
have thought it over, and mean to have a
try.   ·

The speaker, confident in his own persuasive powers, persists, but interruptions, hoot-
ings, and hissings become first frequent, then
incessant, and missiles begin to be thrown,
till at length the great statesman is smug-
gled out the back way under an escort from
the leaders of the disaffected.

In a twinkling the same crowd appears
on the open field. It has been transformed,
many being in uniform and armed — many
more provided with such rustic weapons as
they could lay hold on. They have leaders,

and are being drilled. Even if the police had not melted away, nothing but regular troops could touch so great a force. As we come on the scene, a force of regular troops is advancing. Their leader, having ascertained the position and numbers of the enemy, has formed a plan of attack and instructed his officers. The insurgents, cheered on by their European and other leaders, stand their ground, and have begun to open fire. A few bullets whistle through the air, and one soldier even is wounded. The officers fall behind, and order the troops to advance. This draws murmurs from the soldiers, who demand to be led. Seeing the hesitation, the insurgent leaders advance their men, and strengthen their fire. One or two men fall. This is too much for the student officers, who end by fairly turning tail and bolting, followed by the jeers and curses of their men. The insurgent leaders, watching through their field-glasses, send an envoy to the troops with overtures, and a negotiation is begun which ends in the troops joining the banner of the rebel chief.

My dreams end here, but it is easy to guess that, by much suffering, heavy losses, and fierce struggles, this lesson was burnt into the minds of the people of Bengal — "Deliver not the tasks of might to weakness."

QUARRELS

QUARRELS.

THERE is one more dream about my native neighbours which I will record, and that relates to their quarrels. The importance of a quarrel as showing character is that there is more reality, more sincerity—less disguise, pretence, and humbug—about a man when he is engaged in a serious quarrel than at any other time.

By quarrel I mean a contest between two parties in which, if the one side wins, the other must lose. The parties may be face to face; or the one may pursue and the other flee; or one may be plotting against the other, who is unconscious of it. The quarrel may be between parties of any numbers, from nations to individuals; it may be sudden and soon over, or last for many years; it may be on one spot, or spread over a large area; a simple event, or a game of many moves and pieces. In all quarrels there is something to be lost and won, and it is a matter of great interest to watch what are generally the stakes, and how the game is played.

So far as my native neighbours are concerned, quarrels on a great scale—national quarrels or wars — are things of the past. The invader no longer makes raids from outside India; nor are the races within the country, such as the Mahrattas, allowed to attack their neighbours. Fighting and wars there still are; but they have been taken over as its own affair by the British Government, and I doubt if one man out of ten thousand of my native neighbours gives them a thought. This immunity from war has brought, no doubt, blessings in its train—not unmixed with the reverse. For one thing,

it is accepted by these people as if courage
and forethought had nothing to do with it,
—as if it were a gift of nature. Again, it
has reduced the capacity of the people for
self-defence; and lastly, it has driven the
instinct for quarrelling which all men have
inwards, and removed the very powerful
bond of union which successful resistance
to external attacks gives to a people.

We hear of the fabled Arcadia where all
men are loving and neighbourly. It is a land
that was never seen. Men will quarrel—if
not about one thing, then about another.

After war on a great scale ceased to be a
concern of my Indian neighbours, broils and
disturbances, little short of war, survived in
the land for some generations.

The wild tribes used to issue from their
jungle fastnesses and raid the more peaceful
plains; the dacoits, or gangs of robbers, har-
ried with open violence the houses of the rich,
and the goods of merchants on road and river,

sometimes on land, sometimes afloat; the great landlords sometimes turned out their little armies of ryots and clubmen hundreds strong, to do battle, not without bloodshed, for disputed territory; or leagues of ryots combined to resist by force the claims of the landlords.

The wild tribes have for the most part been pacified. It was found that their wildness was caused as much as anything by their being treated like wild beasts. Take, for example, the Rajmehal hillmen. During the eighteenth century they were dealt with as banditti, and picketed round with colonies of fighting men, whose business—very badly performed—was to hem them in and stop them from raiding. So badly was this business done, that Government superseded the landlords, who were in charge of it, and sent in troops. The officers won the confidence of the hillmen, and finally these tribes were organised; a regiment of militia was raised among them to keep order, and a court of their chiefs established to try offenders against justice. I remember reading an account of a trial before this court of over 100 hillmen, who had made a raid in which many murders were committed and about 900 head of cattle lifted. While the trial was going on many of the prisoners had to be made over for custody to their chiefs, as whole villages had been left without a man to look after the women, children, and stock. It is about eighty years since we have heard of such occurrences. Hillmen are still fond of stealing cattle, but they no longer do so openly or with violence.

Then there was the Sonthal rising in 1855. Driven to desperation, as it afterwards appeared, by the exactions of the money-lenders and the police, they took to arms, and had to be put down by an armed force. As with the hillmen, so with these, the method adopted was to gain their confidence and organise them, protecting them as far as possible from oppression in the name of justice. The result has been that they are a loyal and contented though high-spirited people. They have to be watched and managed, but no longer break out into violence.

I remember reading accounts of the supression about forty years ago of the gang-robbers who used to infest the banks of the Hooghly, plundering the traffic which by land and water passed in large volumes to and from Calcutta, and terrorising the well-to-do inhabitants of its environs. Gangs of fifty or sixty at a time were brought before the courts and tried. Many were convicted and sentenced to transportation for life. The places where they used to lie in wait are now occupied by busy jute-mills. I remember in particular the story of one man who was charged with being a notorious and desperate robber. His defence was that he was a broken-down and feeble old man, incapable of the active deeds ascribed to him. On his appearance, this was believed, and an order was passed for his discharge. As soon as he heard the order, with great imprudence he turned a back somersault out of the dock. His premature glee cost him his liberty, as he was brought back and sentenced. There are still gang-robberies. Gangs form wherever they get an opening, and doubtless a year or two of neglect would cause a tremendous growth in them. But they are no longer the sturdy and desperate bands that once swarmed over the country, and as soon as a gang becomes formidable it is detected and broken up.

What we call a river in India is more like what in England would be called an inland sea. In the height of the rainy season the Ganges spreads out to a width of twenty miles. Its delta is a network of deep and gloomy creeks bordered by dense forests, and very lonely. Even to-day the forest is extensive, though immense tracts have been cleared. In the islands of this delta the pirates—Portu-

guese and natives—had their nests, whence
they were rooted out by the Mogul princes,
and here their descendants in our time moved
about in swift boats, plundering richly-laden
vessels, and sometimes also villages on shore.
A few may still escape the vigilance of the
police, but the energy that was once expended
on fighting and robbery is now for the most
part diverted into the more honest industries
of fishing, and serving as lascars on river and
ocean steamers.

When landlords disputed the ownership of
land—it might be a new island found at the
end of the floods to have been formed in
the bed of a great river, or it might be waste
land of any kind—many were the moves in
the struggle. Men would go with boats or
carts to cut whatever produce there was—
trees, grass, or anything—to stick up sheds,
to dry fishing-nets, or perform any other act
of ownership they could think of; grazing
cattle, ploughing, &c. A rival party would

arrive from the other side with similar intention, and pitched battles often ensued. For
a newly - formed island, though next season
might see it swept away again, might also
become a rich estate, thousands of acres in
area, of good land.

Again, according to the law of the land, an
estate may be held jointly by a number of
partners, each of whom has certain rights
over it, and may sell, or mortgage, or sublet
his or her rights. There used to be notorious
speculators, with disciplined bands of retainers,
who would buy up a tiny share of an estate, and
then by sheer force prevent the other owners
from collecting rent, or making any other use
of the property — until they were ready to
part with their interests for an old song.

In all quarrels about land, the great thing
is to get possession, and so drive the opposite
side, if it desired to enforce its claims by law,
to be the plaintiff, on whom ordinarily lies
the burden of proof.

When fighting was to be done, the local
men managed their battles, aided sometimes
by hired bravos. The hired bravo was often
distinguished by the blindest and most devoted courage, and absolutely reckless in
exposing his life to danger. A story is told
of a landlord in Eastern Bengal who was
out with a solitary attendant, and saw advancing on him a large band of the enemy's
clubmen. "Run, sahib," said his henchman,
"and I will meet them." "But they will
kill you," said the master. "Very likely,"
said the man, "but never mind. What a
splendid case you will have in court!" This
is the stuff of which professional clubmen
were made.

The fighting qualities of local people varied
greatly. There were the industrious but ef-
feminate Hindoos of the plains around Calcutta, whom one or two bravos would suffice
to overawe; who, to save their skin, would
yield to almost any demand. Then there

were the manly Rajpoots, Ahirs, and Puritan Mussulmans of Behar, dwelling by the Ganges, and wielding the great iron-bound club, which they whirled round their head, and one blow from which would dash out a man's brains. There were the Gwallas of Nuddea, also renowned wielders of clubs. There were the sturdy Chandals of Fureedpore, described by tradition as a whole community of Hindoos, outcasted from Bramins downward, who made themselves homes in the swamps of Fureedpore, raising large mounds to build their houses on, and living chiefly on fish. Their favourite weapon was the spear. Then there were people lower down the delta of the Ganges — Puritan Mussulmans of Backergunj, who fight with guns, and their coreligionists in the districts of Tipperah and Noahkolly, still farther east, where the ryots are, man for man, a match for an equal number of clubmen. Then, too, there are the wild tribes of the western

jungles—the Sonthals, Kols, Bhuyas, Paharias, and others, with their bows and arrows
and battle-axes ; and those of the east and
north—as the Dufflas, Nagas, and Bhuteas,
the Kookies, Chukmas, and Looshais, with
their *dahs* and spears.

Here and there we catch a glimpse of
the stormy old times, now little more than
memories.

Tradition tells of the great indigo-planter
who employed large bands of devoted followers, and was so well served that he, a
private person, was able to pass a sentence
of twenty-five years' imprisonment, and carry
it out, in spite of the law. It is now many
years since his factories, whose ruins may
still be seen, were stormed and destroyed by
the peasantry. There is a story of another
landlord, who, when settling his rents with
his Mussulman tenantry, had a basketful of
their beards brought in every day till the
tenants agreed to his terms. They tell of a

rajah whose tenants, in British territory, declared that his rent-collector had defrauded them by taking their rents without giving receipts, and again suing them in the civil court for the same rents. The tenants made a league, drove off the rajah's servants, the process peons of the courts, and the police, issued a no-rent manifesto, and even invaded the rajah's independent territory, destroying a small fort, and retiring with the plunder.

The causes of strife varied from district to district.

The tension and strain in the closely-packed, over-crowded populations of Behar, where men were struggling for land, were very different from the conditions in the thinly-settled tracts of the eastern, northern, and south-western jungles, where land was lying waiting for men to cultivate it. Here landlords crushed tenants; there tenants defied landlords to exercise their rights. Sometimes race pressed upon race; sometimes the middleman

squeezed in between landlord and tenant, or
the money-lender sucked them dry, or, wax-
ing fat with prosperity, men grew proud and
fought for glory or jealousy. But seldom was
there peace.

Fifty years ago, with few roads, slow communications, a scanty staff of Government
officers, absence of capital, and little or no
trade, the great landlord was a king in fact,
whose writ ran, whatever might be the theory,
before that of the Crown. He controlled the
police, and held all the local forces in his
hands. Except in a very few small areas
to which his authority did not reach, he was
king in his estate, and the local people were
his subjects, though not always very orderly
and obedient subjects.

By gradual advances the law has strengthened its influence, and brought check after
check to bear on the exercise of physical force
by subjects.

Great and constant efforts were made to

induce the people to come to the law courts, instead of resorting to physical force, or the cutcherry of the zemindar. This is one secret of the notorious and somewhat scandalous tenderness shown by our courts to false witnesses and perjury. Harshness, it was feared, would frighten people away altogether.

The police were weak in numbers and of doubtful reputation. A story I was once told by an old indigo-planter is a fair illustration of what was thought of them. A friend of his had a case which the inspector, or Darogah, as he was called, came to investigate. When the inquiry was over, the Darogah produced three draft reports, the first of which he said he would send on for 500 rupees, the second for 50, and the third without charge. As the first was likely to be the cheapest in the end, the 500 rupees were paid. In the great quarrels, the Darogah often used his influence on one side or the other, or was paid heavily by both to be neutral.

The old police were reorganised and strengthened over thirty years ago. The courts, too,
have been greatly increased in number and
strength.

An important step towards peace was the
Revenue Survey, which, by fixing the boundaries and recording the ownership of land,
made it less profitable to fight over these
matters.

Then there was the Arms Act, restricting
the use of arms, and compelling those who
had them to take out licences.

Then the criminal law made a landlord
amenable for a riot committed in his interests.
Before that, he had been able to sit safely
at home, and send out his retainers to run
all the risks. He was also now compelled
to deposit lists of his retainers with the
magistrate.

Again, the laws of procedure enabled a
magistrate to award possession of disputed
land to one of two disputants, and to bind

over persons who were likely to break the peace.

Courts were multiplied. I know of two cases in one part of the country where new headquarters of magistrates were established —the one to restrain an unruly peasantry, the other to keep in order a truculent family of landlords.

Police stations were established wherever the condition of things seemed to require it, and special police were quartered in disturbed neighbourhoods at the cost of the inhabitants.

In fine, the Government and its officers came to have more to say, and the law was more difficult to defy.

Breaches of the peace, violence, and riots have not ceased to occur, but they are now less numerous and on a smaller scale. Whether this improvement is permanent or not is doubtful. The peace is kept, not because to keep it is natural, but by the constant use of watchfulness, and application of moderate but ample and resistless force where and when needed for its preservation. On the Assam frontier, in the hills and forests overhanging Orissa, in the ancient state of Manipore, among the religious fanatics of the Ganges valley, and in the streets of our great towns, we have from time to time warnings by disturbances to be on our guard, and that there are forces ready to break loose if they get the chance.

To keep those forces weak and under control, two means must be used. The first is good government, by which great grievances will not be allowed to spring up; and the second is strength used to put down any defiance. To the froward we must be froward. If the master displays weakness he will be sprung upon and rent. There is one quality which is general among the people of Bengal. If they are strong, they show no mercy. If they get an enemy down, woe to him! There is nothing they respect like power — not even justice or kindness. " Of

all the *hakims* that have been in this place,"
a man was asked, " who was the greatest ? "
" There have been many great *hakims* here,"
was the answer, " but the greatest of all was
P—— Sahib." " And what did P—— Sahib
do more than all other *hakims?* " he was
asked. " He put a rajah in prison," was the
reply. " But other *hakims* have put rajahs
in prison, when they deserved it," was ob-
jected. " Ah! but this one did not deserve
it." They have not yet got beyond that
stage of civilisation in which power is the
cardinal necessity for a ruler; and we must
not forget this, because our own power is
being used more in the interests of our subjects than has been the rule among our pre-
decessors. We may not draw cheques so
frequently, but we must not part with or
reduce our balance at the bank.

Coming now to more recent times, I should
attribute any difference between the present
day and those more remote to two main

causes — one being the success of the efforts
made by the British Government to increase
the influence of its law courts ; and the other,
the growth of trade, with its accompaniments,
easy communication, education, and diffused
wealth. We must not run away with the
notion that these two influences have reached
the same position, or have the same effect in
India, as the corresponding influences in England. Blind and feeble are the eye and hand
of the law; small and trickling the streams
of commerce in India judged by the English
standard, and that in India is looked upon as
wealth which in England would be despised as
poverty.

Still progress has been made, and has modified, whether for the better or not, the old
methods of quarrelling. Progress has been
greatest at the centres of trade and communications, where are the seats of authority, and
where dwell most of those who think, speak,
and write. Naturally, therefore, there is a

tendency to exaggerate the amount of progress actually made.

One who is a member, as a dweller in a large town is, of a compact, well-protected, and closely - governed community, is apt to think little of the vast and monotonous expanses, overspread with thousands and thousands of small communities detached from one another, isolated from the central authorities, each a little world to itself. He is apt to give the people in them about as much thought as the human inhabitant of the village gives to the animals in the jungle near which he lives, or the fish in the river that flows by his house. When they occasionally put out their snout and become visible for a moment, his eye may be caught, but he seldom follows them in imagination back into their haunts, or pictures to himself their anxieties, difficulties, perils, or escapes, their stratagems, their friends or their foes. If he does, then he peoples their wilds with the institutions and characters familiar to him, as Uncle Remus did when he dubbed the bear a judge.

Towards this world, with which, peopled by Hodge in his millions, the reader has been made acquainted, have approached the law courts and police ; the railway, the post-office, the telegraph, and the school. Hodge and those he has to do with must give heed to those new factors, and does his best to understand what they mean.

Passing by all of these modern influences except the law courts and police, we have now to consider how a quarrel is carried on within their "sphere of influence." This expression seems to me a very good way of describing the relation of the people to the law courts. Their hold is the strong but loose hold of the British in the regions of Central Africa, rather than the tighter and more certain grip of the Government in England.

The primitive method of quarrelling is the crime of Cain — the taking of human life. In a country like India, where for countless ages human beings have swarmed in their millions, the life of a man is not the thing of importance that it has been for many centuries past in England. In India importance has always, in the native mind, been attached to the lives of Bramins and men of the higher castes, but not to common clay. A man does not seem to value even his own life as much, or to struggle for it with the same determination, in India as in England. He who has taken life in that country is seldom troubled with the remorse of Eugene Aram or Jonas Chuzzlewit. Two things only he believes he has to fear — the vengeance of the victim's relatives, and the leaden-footed pursuit of the law.

Although the law does occasionally catch up transgressors in this direction, still, except when the friends of the victim are in-

R

fluential, determined, and prompt, and the
taker of his life is neither himself powerful
nor under powerful protection, there is not
often even a serious pursuit, nor does the bare
fact always come to the knowledge of the
authorities. There are no local magistrates;
no coroners to hold inquests, or juries to
check proceedings; no doctors to certify the
cause of death. The heat of the climate
renders necessary prompt disposal of the
dead, usually by fire. Cholera and snake-
bite are convenient causes to assign for any
sudden death. Dead bodies of persons who
are said to have been murdered are taken
to the one qualified medical man in the
district—the Government surgeon—generally
in a state of decomposition, and he examines
them in order to detect the cause of death.
But in so few cases do the bodies reach the
surgeon that his post - mortem examination
is probably of more value as a protection
against false accusations than as a means

of bringing to light the guilt of actual crim-
inals.

In addition to the ordinary weapons that
naturally lend themselves to the lust for
taking life—the deadly weapon, the common
household instrument, the club or stone—is
poison. In India there are no chemists'
shops with diplomas, labels, registers, and
certificates. Arsenic is kept in most houses,
and can be bought in all bazars, while many
kinds of deadly drugs are to be had for the
gathering in the jungles and waste land
that are near every village.

To give as vivid a picture in brief as I can
of the varieties of death inflicted and the
motives that lead up to the tragedy, I will
endeavour to call up a few of the hundreds
of cases that have come within my own
experience, and present them to the reader.

I see on the side of the village lane three
small huts round a tiny courtyard—a square
of 2 or 3 yards. The huts are inhabited

by three brothers, each with his wife and family. One of them, the worker of the family, has long grumbled because his brothers will not help him to plough, sow, and weed their common fields, and has at length enraged his eldest brother, the idler, by threatening to separate, and even go to law. Day by day the wrangling is renewed, till one night the eldest, after much brooding, silences his brother and prevents the lawsuit, by cutting him up with an adze as he lies asleep in the open verandah.

I see in Eastern Bengal a flat expanse of rice-fields laid out in small squares, which are separated by little mud-ridges covered with grass. One morning Jan Mahamad goes out and finds that his neighbour, Ahmad Ali, has pared away the turf on his own side of the boundary ridge. This has by ever so little moved the centre of the ridge towards Jan Mahamad's field, and so enlarged Ahmad Ali's, whereon arises a controversy, which does not end till Ahmad Ali's lifeless body is found one day in his field, the head split open with a hoe.

Again, the country is parched with drought, and there is a pond, a precious reservoir containing the last few drops of water with which may be saved a small part of the perishing crops. Jealously has this been guarded, and carefully has the share of each been apportioned. But Nobin Moudle, who is on the watch, catches Ali Bux in the act of drawing off water out of his turn. He seizes the hoe with which Ali Bux is clearing the channel, and drives him off. Ali Bux, furious at being balked, steals that night into Nobin's house, and cuts him up with a chopper on his bed. Nobin's wife, who is sleeping just inside the open door, wakes up and sees this going on in the verandah outside by the clear moonlight.

I see a Sonthal march up to the house of a British magistrate with a bundle under his mantle. He throws down the contents — a human head—and says, "I found a man assaulting my wife, and cut off his head. Here it is. Do with me what you will."

I see going on in a Mahomedan village of Eastern Bengal a grave consultation among heads of families. They are discussing the conduct of a dissipated young man who is a trouble and a danger to the peace of their households. It is not a matter that they care should go into the publicity of the courts. He is condemned, and a few nights later is waylaid, while prowling as usual, on a lonely roadside where the bamboos are massed thick overhead, and taken into the jungle. There he is held down by some, while others pound him to death by driving their elbows into his body. In Eastern Bengal they generally use this way of beating and murdering. It can break the bones, yet leaves no mark.

I see a consultation of substantial peasants in a Midnapore village. Among the Hindoos it is the custom to cast out the dead bodies of

cattle in a place where they are first stripped
of their hides by the local *moochi*, or worker
in leather, and then disposed of by jackals
and vultures. The *moochi* in this village has
been artificially adding to his tale of hides by
poisoning the village cattle. This practice
reached such a pitch in one district I knew,
that a general order was issued to slash the
hides of all cattle that died, and by making
them useless remove the motive for poisoning
the cattle. Twice has the man been caught
and sent up for trial, but both times he came
back triumphant, acquitted. The neighbours
remonstrated, but he would not leave off his
bad practices. The consultation is now being
held as to whether there is any further
remedy. There is. They catch the man and
lynch him. I was the magistrate before
whom these men were brought up charged
with his murder, and this was the story which
they told, simply. "What were we to do?"
they said. "We could not see our cattle

being poisoned day by day. We tried to get him to leave off, but he would not heed us. We asked the law courts for help, and they would not give it. We had to kill him." Their powerful statement of their dilemma affected me more almost than anything in my experience, and has impressed me strongly with the feeling that, if Judge Lynch is to be kept out of the field, Judge Law must do his duty. These poor men were sent for trial to the sessions, and I rather think they were hanged; but I pitied them with all my heart.

I see a household of Sonthals. The Sonthals believe in witchcraft, and attribute sickness to its spells. There are two households, that of the elder wife and that of the younger, in both of which there are married sons. The father lives with the younger wife, who is, as usual, his favourite. In the house of the elder wife there is mourning, for her son has just lost his little child by death. He sends his younger brother to summon his stepmother from her house, which is about 50 yards away. She comes weeping aloud in response to the cries and wails of the household, and is met at the door by her stepson, who knocks her down and beats her to death with a heavy club. Then he rushes off after his own father, who flees through house after house, and at last finds refuge in a field of standing crop. When put on his trial the son said : " Of course I killed her. She ate up my child by witch-craft, and it was not the first." His father had taken her side.

Another Southal I remember. He lives with his father and four children. Three of his children die, and he and his father go to the witch-doctor to find out why, or rather who, it was. The witch-doctor says his next-door neighbour has a demon in the house, which is doing all the mischief. They return, and a village assembly is called, before which the neighbour is solemnly warned to take better

care of his demon. Of course he denies that
he has such an inmate. Soon after, the father
and the remaining child die, and the man,
bereft of all, is fully persuaded that it is all
owing to his neighbour's demon. He feels his
duty clear. At a ceremonial feast in a house
in the village, at which the culprit is a guest,
he is helping to distribute the food, and takes
his neighbour's portion to him. On the way
he puts poison in it. Being on his trial for
this, he said, " Of course I did it. I don't see
how I could have helped."

I remember one remarkable instance of an
attempt which failed. It was a man who
sneaked into a neighbour's courtyard to put
poison in the water-pots. He was surrounded
by the dogs of the house, which not only kept
him from reaching the water-pots, but also
prevented him from getting away till he was
caught with the poison by him. I have
thought better of pariah dogs ever since I
heard of this.

These are a few—all facts—illustrations of what goes on all over Bengal. The number and variety are very great, and the few specimens here given can convey but a dim notion of the tragedies, unrecorded yet often exciting and dramatic, which are being enacted.

The more commonplace methods of carrying on a quarrel need less detailed notice. Perhaps the most curious is that adopted chiefly by young women, introduced too early in life to the cares of the family, who, to be revenged on the husband, mother-in-law, or elder wife, rush off after a scolding-match and drown or hang themselves.

A characteristic method of quarrelling is the social method. There is not much need of explaining in these days what the "boycott" means. When he has been sent into Coventry, a native cannot get any one to eat with him, shave him, wash his clothes, or conduct his family worship. His children cannot

get husbands or wives. He is left severely
alone.

This is the penalty inflicted by the caste on
one who has been denounced to it for a fault,
and has failed to clear himself. Used in a
legitimate way, it is the instrument by which
caste discipline is maintained. The usual way
in which it is received is by making submission, giving up the practice condemned, and
paying a fine or giving a feast, as custom requires. Sometimes, however—whether from
contumacy, or from conscious innocence, or
from the knowledge that the real motive of
the punishment is not the good of the community but the gratification of private spite
or ambition—submission is withheld. In a
system where the "boss" has his place, there
is a tendency to substitute his will for the
lawful customs of the caste or society, and to
punish with its penalties disobedience of that
will. Most disputes either begin with or pass
into the stage of the caste, the "boss" seek-

ing to use his majority for putting down his own or his followers' enemies, and the opposite party supporting them. If the quarrel lasts long enough, and the sides are both influential enough, the caste splits up, and the opposition comes to have its own organisation, as, to compare small things with great, the Liberal Unionists have done in England. Each has its priest, barber, washerman, and is complete in itself. Splits of this kind will last sometimes for a generation, and enmity will be active. If the disputants are great men, there are many whose interest it is to keep the quarrel alive, and that for two reasons. The first of these is that a powerful family, if united, is able to oppress its neighbours, whereas, if it is divided, the members are fully occupied with their family quarrel, and both sides try to gain strength by courting popularity.

The second reason is that, when a quarrel is going on, there is a great deal of expen-

diture that cannot be audited, and affords
good pickings to lawyers and servants. If,
for instance, large sums are entered in the
accounts as disbursed in bribes, the employer
has to take the word of his agent that it
was actually paid.

The disputes of the great, especially between near relations, last long, spread wide,
and are fought out with great determination.
I know of one celebrated faction dispute
between brothers, two prominent members
of society, which lasted thirty years. The
brothers lived within sight of each other,
but never spoke or had any intercourse, and
the multitude of hostile acts done on both
sides was beyond count. The disputes of
the poorer classes do not spread so far or last
so long, as there is not the same interest for
outsiders in fanning them, but even they
often last for years.

While a social dispute is going on, a
continuous effort is being made by bribes,

entertainments, promises, threats, and even violence, to detach people from one faction and add them to the other.

Of minor methods of annoyance, the following are some : Setting fire to the house; driving cattle through the crops; letting loose and pounding a man's cattle; destroying crops; throwing stones or filth into the homestead; using bad language in hearing of the women; insulting the women on the way to the well or the bathing-ghaut; insolence to and assaults on the men; sending round defamatory rumours against the character of a family; if the quarrel is about land, taking forcible possession of it.

It may be said that many of these methods of annoyance are in use in our own country. Here, however, whole communities co-operate to do these things, get them done, and screen the perpetrators.

Many of these acts are punishable by law, and up to this time no mention has been made of the law and police. We have now to examine to what extent these have influenced the quarrels of the people.

In a district where there are strong men of influence, little trouble is given to the law courts by the public. If the man of influence is a just man, they care not to go— if he be oppressive, they dare not go — to court. For in the one case redress is easier to get; in the other, it is only to be got elsewhere and otherwise.

There are two great divisions of the law courts,—the civil courts, for deciding matters between private persons; and the criminal, for deciding quarrels between the State and its subjects.

In case of a civil dispute, the great object of each side is to drive the opposite side into court as plaintiff, thus gaining the advantage of possession. One of the most important points in a civil suit is that of the burden of proof, which means that if neither side can prove its case, the side on which the burden of proof rests must lose. That side is generally the plaintiff; therefore it is an advantage eagerly sought after to enter the court, not as plaintiff, but as defendant. The questions in which this advantage is of the greatest importance are those of right to land; and naturally, unless a claimant is in possession, he must go into court before he can exercise any right. Long and desperate are often the struggles before either side will admit that it is out of possession. At ploughing-time and at harvest-time strong forces of ploughmen or sicklemen, as the case may be, sally forth, attended by guards of clubmen, to plough up the land or reap the crop. The opposite side sally forth to drive them off, if strong enough ; or if not, to stand at a distance and call on the Magistrate, the Company, and the Great Queen to put a stop to this great wrong. Sometimes the controversy does not go beyond words; but it is often decided, especially on the banks of the Ganges,—in Behar, or in its delta, Backergunj or Fureedpore,—not without loss of life, by the weapon of the country—the iron-bound club, the spear, or the gun.

Quarrels about land are on all scales, from the petty squabble for a single field, or a scarcely visible strip on the boundary, to a great struggle involving many thousands of acres. In whatever way possession may be taken, about that quarrels arise,—about catching fish; cutting wood or grass, gathering fruit or leaves; quarrying or collecting stone or lime, mining; putting up houses, grazing cattle, mooring boats; establishing markets. One side tries to do the acts, the other to prevent its doing them. When the land is cultivated, each side brings pressure to bear on the cultivators to pay rent solely to itself. Fighting about such matters frequently goes a long way without any one being punished, owing to the dislike felt by the law courts for punish-

ing men who are merely defending their rights,
which both sides of course claim to be doing.
If it goes too far, however, the magistrate
generally steps in and settles the matter by
declaring one side or the other to be in possession.

At last the opponents meet in the court,
and get to close grips for the final tussle. It
has been anticipated and prepared for long
beforehand. In the old days the great landlord had, somewhere on his premises, a manu-
factory of forged title-deeds, bonds, and other
papers. Stamped paper of all ages was
bought and stored, and skilled artificers were
employed to imitate handwriting, and give
the documents an appearance of age. This
trade, as appears from the proceedings of
the law courts, is not yet extinct, though
the Registration Laws have reduced its
dimensions.

The contending parties in most lawsuits are
men of small or moderate means, but a man

will usually spend all he has and all he can borrow in fighting his case.

In the law court, as outside, the long purse gives its owner an immense advantage. It can pay good lawyers; bribe the confidential servants of the other side to give useful information or papers; cool their zeal, and get them to make mistakes. It can buy and bring forward witnesses on his own side, and frighten or corrupt those on the other; and it can forward his business in the court by timely gifts to the various officials about it. When the case has been decided, if he loses he can afford to renew the battle again and again with applications of appeal, for revision or for review, and so wear down his opponents. Those who have more friends than money depend on the work outside the court, frightening away the evidence of the opposite side, and marshalling hosts of witnesses on their own.

After the decree is fairly established, and

cannot be shaken, comes the difficult task of executing it. A decree is nothing till it is executed, and that process is delayed by evasion, violence, fraud, corruption, and all manner of means. The struggle, before it ends, is often one over the costs, which have probably mounted up far above the intrinsic value of the property fought over. England is not the only country where the lawyers get the oyster and their clients the shell.

But is it not stupid to go on fighting—spending much for a comparatively worthless object? Many people say it is, and laugh at those who do it as foolishly litigious. But many as are the motives for which a lawsuit is entered on, fun is seldom one of them. In Eastern Bengal, indeed, litigation seems to be a popular sport. It is said that many Chittagong men, who have made a modest fortune at sea, look out, when the time comes for them to settle down on shore, for a nice piece of land with a good tough lawsuit on it. In Tipperah the youth who has won his first case is as proud as the English lad who has made a hundred runs at cricket.

In general, however, the mood of the litigant is more serious, as well it may be. He may have been dragged into court, for at least one party is there against his will. A man goes to law, or accepts a challenge, for his own safety and comfort. He has many business relations with many men, and their treatment of him will depend a good deal on how he defends his rights. If men see he will fight resolutely, they leave him alone. "If you wish for peace, be ready for war." Many struggles between determined opponents end in a compromise, and this more especially if they find some one of such rank that it will be no disgrace to do his bidding. Thus the knight of old fought on, and sometimes gave up his life unless he found one to whom he could honourably yield his sword. I remember many instances in which I and

others were able to settle old - standing
quarrels; and this is one of the chief, and,
when successfully performed, most pleasant
duties of British officers in India.

As for criminal cases, they are fought out
more sharply. There is never any question
of burden of proof in these, as that is always,
in British courts, on the prosecution. One
accustomed to British ways will find it hard
to realise those of Bengal.

We are to suppose that the matter is to
go to court, and has not been suppressed or
hushed up. There is a rush to the police or
the magistrate, and a wild mixture of plausible false charges with possibly some that are
true. Liberal use is made of "embroidery"
—twenty men being charged where the real
case is against one, and serious charges hung
round the neck of the comparatively frivolous
fact that is true. Most probably the false
accuser gets the start of the true.

The case gets into the hands of touts, who,

to use a figure, are splendid skirmishers and foragers, and skilful in taking advantage of cover. Criminal charges are often made use of purely as a weapon of annoyance against the enemy. Thus if a man's house is burnt down by accident, he promptly takes advantage of his misfortune to lay a charge of arson against his enemy, and has ready numerous witnesses who "happened to be out at the time" (midnight), and saw the match applied. Or if he was an unsuccessful suitor for a woman's hand in marriage, he will prosecute her for bigamy when she marries, and have plenty of witnesses who were present when she was previously married to him.

Or, if some of his people meet with an accidental death, a neighbour is charged with the murder. I remember a family quarrel in which there were a father and two sons— the younger favoured, the elder hated. The younger hid away, and the father prosecuted the elder son for the kidnapping and murder of his brother. That charge was proved false, and the father was prosecuted for making it. Soon after, the father, while out on bail, having had a quarrel with the elder son in the fields, fell down in a fit and died. The younger son then prosecuted his elder brother for the murder of their father, and had witnesses to prove that he was clubbed. The evidence of the surgeon showed that to be false; and a fresh set of witnesses were produced to prove that the father was strangled by the son, which also the surgeon pronounced false. Corsicans are not more vindictive.

I have attempted in the next chapter to give a livelier notion than any general description can, of the infinite variety of intrigue and rascality that goes on in our courts.

A Service poet has aptly described the occupation of the Indian judge as

"To dive in wells of perjury for truth."

It is more like groping in quagmires of perjury for the same. Partly owing to the lenient attitude which, with the hope of drawing people to our courts, the Government and the Legislature have adopted towards perjury; partly because falsehood and fraud are common in the country, the commission of perjury in court is not seriously condemned by public opinion. It is rarely punished by the law, because, to punish it in a criminal court, the burden of proof must be discharged, and proof depends on evidence between which and that which is to be proved false there is seldom much to choose. Yet on evidence which is not good enough to prove perjury, heavy and important claims are decided; and men are condemned to death, transportation, imprisonment, and fine.

This is the weak point—the doubtful good of the law. The courts are beset by hired champions of all grades, from the chivalrous barrister and high - minded pleader, who are rarely let into the secrets of the trickery used, down to the unscrupulous tout, who tutors false witnesses, flings about bribes and threats, and generally does all the dirty work for his side. Perjury is opposed to perjury; false charge to false charge; handfuls of dust are thrown in the eyes of Justice, first by one side, then by the other, until she is in danger of becoming blind in reality, and is too often blear-eyed.

Was it well to withdraw the supremacy from the champions of physical force, and hand it over to those of the tongue and the pen? Physical force still reigns, for the Government wielded and wields the power which controls other powers. But formerly the great landlords and men of wealth were not much interfered with, and spent their wealth on the maintenance of bravos. Now the power of the State is placed at the disposal of the courts, and the courts are swayed by the paid lawyers, witnesses, and other means which the wealth is used to provide instead of the clubmen of old. Let us hope that at least the change is not for the worse.

Skilled judges of probabilities often mingle physical violence with legal proceedings, as did the litigants who caught their opponent at the back of the court-house and gave him a great thrashing, rightly judging that the court would not believe his story, but would look on it as a trick.

The whole business is a game of odds and evens, where my constant effort is to guess which of two things you have in your hand, and you are trying to do the same by me. Only two alternatives, and yet an endless mystery. So ends my dream of quarrels.

POSSESSION

A DREAM OF LAW COURTS

POSSESSION.

A DREAM OF LAW COURTS.

ONE hot afternoon in May, about twenty years ago, there was excitement in the village of Fureedpore. It was not the sort of year when people are easily excited, for there had been a good fall of rain, and the people were busied with their ploughing.

The fact was that Hurbullub Sing, the old hereditary landlord of the village, after a long struggle with debt and difficulties, had at last failed to pay the Government revenue. His title had been put up to auction, and bought for 15,000 rupees on account of a Calcutta merchant, one Rámchandra Shaw.

The collector's bailiff had stuck up a bamboo, and proclaimed by beat of drum the change of ownership, and that all dues to the owner must now be paid to Rámchandra Shaw. Rámchandra had appointed as his *naib* or deputy one Rám Ratan Rai, a friend of his, who had come to the village and gained the support of Ratikanta Ghose, a well-to-do ryot, whom he appointed local rent-collector. It was now time to make the *punnia*, and the prospect of this ceremony was the cause of the excitement. Every year the landlord makes his *punnia*—that is, holds a meeting of ryots, and takes from each a small nominal sum, as a sign of acknowledgment that he is landlord. In ordinary times this is a mere formality, but when there is a change of landlords, it becomes a matter of the greatest importance. The new landlord, unless he has been acknowledged as such by the village ryots, has to undertake a tedious and expensive course of litigation

before he can exercise any of the rights he
has acquired. There was good hope that
the ryots would acquiesce in the change, for
they had been on notoriously bad terms
with Hurbullub Sing, who had worsted them,
both by violence in the village and by fraud
and perjury in the courts. His right - hand
man in all these battles, both physical and
legal, had been one Koylesh, a *dallál* or
quack lawyer.

Hurbullub did not intend to give way to
Rámchandra without a fight, and had called
a meeting of the ryots to discuss the situa-
tion. Partly from old habit, partly from
curiosity, they obeyed; and were accordingly
assembled in the great open space before his
house—the only masonry house in the village
—where, under a venerable banyan-tree, had
been held for the last century all the fairs,
meetings, and festivals of the village. It
was a sad meeting. Hurbullub himself was
bent with care and age, and his face was

darkened with gloom. His house — cracked
and overgrown with weeds — corresponded
but too well with his ruined fortunes. The
instinctive respect of generations did not
prevent some of those present from taunting
Hurbullub with his oppressive conduct in
former times—how he had levied contributions; extorted obedience to his will, preventing resort to the law courts, and generally acted the tyrant. He admitted that
he had oppressed, but argued that he had
taken no more than he needed, when he
needed it; that he had spent it all among
the ryots, and that if he prevented them
from going to law it was for their own profit,
since any man who went to law was pretty
sure to be ruined, whether he lost or won.

He then warned the ryots that their new
landlord would be a very different customer,
and would by hook or crook get their rents
squeezed up, so that he would take regularly
far more than their old landlord had taken,

even including the occasional contributions. When asked why he thought this, he pointed out that Rámchandra was a trader and a stranger, and had paid down 15,000 rupees for the village. He would want a return for this investment, and could not get the return unless he raised the rents. This would end in ruin for all.

The ryots, being impressed with this argument, were persuaded to have nothing to do with Rámchandra; and accordingly, when his *naib*, Rám Ratan, appeared, they joined Hurbullub in driving the *naib* and his two peons, or attendants, out of the village. In the scuffle Potit Ghose got wounded on the head. Hurbullub sent for Koylesh, who at once began to organise proceedings. Criminal charges were to be laid against Rám Ratan and his men, and for this purpose Potit's wound was to be utilised. As this foundation was not sufficient, it was determined to add to it by manufacturing a wound; and accordingly Nilu Bagdi, a low-caste drunkard of the village, was hired for a couple of rupees to let himself be wounded for the occasion. Jodu, the ryot appointed to wound him, struck rather harder than he intended, so that Nilu was not only wounded but killed. On seeing this, Koylesh gave a whoop of triumph, since no one cared a straw for the death of a man so low in caste, and straightway made all the arrangements for laying a charge of murder. The police were sent for, and the body was got ready to go for post-mortem examination by the civil surgeon. Potit also was to lay his charge before the police instead of before the magistrate.

We turn now to the lodgings of Baykanta Mookerjea, a pleader in Dowlatpore, the magistrate's headquarters, about twenty miles distant from Fureedpore. Here Rámchandra himself, who had come up from Calcutta to see about his new estate, was discussing

affairs with Baykanta, who explained to him,
much to his chagrin, that it was not at all
certain whether the ryots would accept him as
their landlord, and unless they did he would
have a good deal of trouble and expense
before he touched his rents. Rámchandra,
who was not inclined to believe that after he
had paid for the property there could be anything more to pay, began to see that there
was something in this view of the case; for
as they were talking, Rám Ratan came in,
walking very stiff, to announce that he had
been beaten and driven away by the other
side. While they were yet discussing what
was to be done next, one of Rám Ratan's
men rushed in to announce that a charge of
murder had been instituted, and the body
was being brought in. Baykanta, who was
not at all sure whether the charge was true
or false, declined at this stage to have anything to do with the case, and sent Rámchandra and his men round to a *dallál* or

quack lawyer called Sital. Sital, at once
coming to the point, urged Rám Ratan to
tell him the truth, whether the murder was
really committed or not. This was denied so
earnestly, that Sital was inclined to believe
Rám Ratan innocent. Sital was, however,
far too artful a man to rest satisfied with a
true defence, so he at once devised a false
one, accepting as many of the facts alleged
by the other side as he could. He instructed
Rám Ratan to admit that a man had been
killed by one of his peons, to allege that it
had been done in self-defence, to hide away
the peon, and to charge the other side with
kidnapping or murdering him. He said that
as Rám Ratan had already lost much ground
by not going straight to the police - station
near Fureedpore, he must now lose no time
in going to the magistrate. Rámchandra now
did a thing which gave Sital grave offence;
when he gave Sital 100 rupees for expenses
he asked for a stamped receipt, and requested

that accounts should be kept, which was not Sital's way of doing business.

Rám Ratan went direct to the magistrate's court, and there found Mr Smith, the magistrate, a young Englishman under thirty years of age, sitting at his desk in a hot, stuffy, ill-ventilated room, rank with the odour o a crowd of natives. He had just been going through his correspondence, and, not long before Rám Ratan's arrival, had heard the court sub-inspector read aloud the first information report of the Fureedpore murder case as arranged by Koylesh.

Rám Ratan, being sworn, now duly gave to the best of his ability the counter-charge as arranged for him by Sital. After a few questions had been asked, however, he was arrested by the magistrate, and sent off to the police, who were already inquiring into the murder case.

When Rám Ratan and his escort arrived, the sub-inspector of police was arranging the terms on which he should support Hurbullub's side of the case. The bribe was 200 rupees, and it was to be reported that Hurbullub was still at enmity with the ryots, who were now fighting Rámchandra entirely on their own account. Rám Ratan, who had had no time to communicate with his master, was now permitted to do so, and wrote to him imploring him to come at once with money to satisfy the sub-inspector. Knowing what he was probably writing, Koylesh sent Rámchandra an anonymous warning that Rám Ratan wanted to decoy him into a remote place where he would probably be robbed and maltreated.

Both the letters reached Rámchandra about the same time. The poor merchant was growing very sick of the whole business. When he bought the property he had no idea that he would meet with any difficulties, and now, with the suspicion of ignorance, he was too ready to swallow Koylesh's bait, and to look on Rám Ratan's letter as a trap laid for himself. He ended by making up his mind to do what must in any case be wrong. He would not go himself, nor would he send the money asked for, but he would send Sital, and let him take another 100 rupees. Sital came at his call, and gave him good advice — to go himself, or, if he sent an agent, to give sufficient funds. Rámchandra would not be moved, and finally Sital went with the 100 rupees, convinced that his case was as good as lost. Still, true to his code of honour, he made one desperate effort to gain over the sub-inspector by the offer of a heavy payment, to be made in the future, with a small sum down. The sub-inspector, well knowing that by the time the money became due the case would be out of his hands, scornfully refused the offer.

Sital's honour was now satisfied, and he felt at liberty to make what he could out of the case, which his employer's stinginess and stupidity had lost.

Rám Ratan was indignant with Rámchandra. "Of course I was going to make money," he said, "but out of the ryots, not out of my master—at least, not so soon."

"To be sure," said Hurbullub; "stealing the seed is poor thieving. It is best to sow all the seed, and wait and steal the crop."

There was not one of those assembled who did not despise Rámchandra, and feel indignant at so poor-spirited a fighter entering the lists, and Sital's suggestion to involve him as an accused person was unanimously adopted. "Money," said Sital, "will pour when he is in. We are all made men."

Accordingly next forenoon, while Baykanta, discussing affairs with Rámchandra, was explaining to the merchant that, having employed Sital to do dirty work, he ought not to be surprised if such a man had unclean hands, the sub-inspector arrived, and arrested Rámchandra on a charge of murder. This was a dreadful blow, all the more dreadful for

its suddenness. The merchant's notions about
money underwent a revolution. Whereas he
had grudged every rupee, his one anxiety
now seemed to be to devise new ways of
spending money for bringing about his acquittal. The most expensive counsel were to
be engaged, and, in short, nothing spared.

We now come to the preliminary inquiry
by the magistrate, Mr Smith. In the same
court—stuffier than ever—Mr Smith had been
sitting for several days listening to the evidence of the witnesses, and this was the concluding day. A barrister—the famous Mr
Bullion—had been brought up at great expense—his fee for the day exceeding the
income of Mr Smith for the month; and beside him was a string of pleaders and lawyers
—all engaged with the money of Rámchandra.
On the opposite side sat the court sub-inspector, a small and rather sad-looking man
in a very shabby uniform.

Koylesh's idea of the effect this would have

on Mr Smith's mind was a just one. It prejudiced Mr Smith on behalf of the weaker side, and, without intending it, he showed a decided leaning that way. It is difficult to stand upright with a man pushing on one side only, and so Mr Smith unconsciously leaned. After the eloquence of the barrister had been duly exerted, Mr Smith ended as Koylesh expected he would, in committing Rámchandra and all his people for trial to the Sessions Court. But he would not make the charge one of murder—manslaughter, he thought, was alone established. Koylesh was full of scorn for Sital, who had brought up a barrister and then given him false instructions. Poor Rámchandra was heart-broken. Like most clients, especially when inexperienced, he had been full of admiration for his counsel's speech, and of confidence that no possible answer could be found. He was therefore prostrated with grief when he found that he was not yet

free, and vented his anger on his agents,
on the magistrate, on his opponents, and
on every one but himself. " The magistrate
was bribed," he cried. "No," said Baykanta, " though I have known men to give
large subscriptions to hospitals or schools,
tanks or roads, that the authorities were interested in, and sometimes they had as good
an effect as bribes. It is, however, too late
for any move of that sort now."

Then began Rámchandra to see before him
a vision of his life to come, and the life
that he had lived. He had lived in Calcutta at his ease. Rice and milk, sweet-
meats and dainties, nautches and drives,
and even great dinners—the smiles of the
great, and municipal honours,—all these lay
behind him. What was before him? Five
years or more of prison life,—striped clothing ;
hard labour ; regulation rations ; no tobacco ;
punishment, perhaps the lash ; nay, even the
terrors of the central jail at Alipore, or the

horrors, unknown and therefore appalling, of the Andamans. He shuddered at the thought, and was ready now for Koylesh to play upon.

This Koylesh proceeded to do, secretly abetted by Sital, who still acted as the merchant's adviser. Holding up before Rámchandra the horrid prospect of jail-life, and thrusting it in his eyes every moment that he was inclined to forget it, Koylesh by its means extorted from the unhappy merchant concession after concession, until he was stripped bare of all he had bought and more. He was to remain nominal proprietor of Fureedpore, but was to give Hurbullub a perpetual lease of the whole, receiving only a little more than the revenue payable to Government. That revenue was a very small sum, and the whole of the purchase-money, with this exception, went to Hurbullub, who thus, notwithstanding the proverb, both had his cake and ate it. Then he lost all his own expenses, which were very heavy, and was required to pay down 5000 rupees more as costs to the other side. On these conditions it was agreed that the case at the sessions should break down, and Rámchandra go free. There was a neighbouring proprietor who had bid against Rámchandra at the auction, and was now ready to offer him better terms, but he was not permitted to accept them, and was forced to drive away the man who made the offer.

Poor Rámchandra went off to weep his eyes out in a quiet corner, and Koylesh and Sital retired to get the deeds drafted, and prepare the case for breaking down. Rámchandra had found out Sital's treachery; but far from being ashamed of it, Sital poured ridicule on him for making it necessary by his niggardly conduct.

So at last the deeds were drafted, signed, and registered, and, as arranged, the case came before the Court of Sessions, and ended

in the acquittal of all the accused. Rámchandra could not make out how such a
thing could be managed. Hurbullub said he
was not very sure, but he generally found that
if he desired a man to be ruined or saved by
a legal proceeding, he had only to instruct the
cunning Koylesh, and the thing was done.

"All this," he said, "is new-fangled, and I
have never been clever enough to understand
how it is done. In the days of my youth we
used to fight it out with clubs, which I had
no difficulty in understanding. Now, however, the Government won't allow that to
be done openly, and I have to consult
Koylesh before I can do even that without
getting into trouble. Instead of clubmen in
the field, we now employ lawyers in the court.
It costs more, but I must say it is more
thorough, ruining at least one side, and sometimes both. I don't know how it is managed.
It is managed, however, as you have seen for
yourself."

" Yes, I see," said Rámchandra ; " but surely all this is very wrong. People in Calcutta will hardly believe me when I tell them of the violent people, the fraudulent lawyers, and the corrupt police that are to be met with in the Mofussil."

" Steady there !" said Hurbullub. " If we are violent and crooked, it is only when we don't get leave to be straight. You, above all men, have no right to say a word, for you came to fight, and cannot complain if you are hit back."

" I had no intention," said Rámchandra, " of hurting any one."

" Oh, a fig for your intentions !" said Hurbullub. " You came to turn me out of my ancestral property ; and if you thought that would not hurt, you were a fool."

" And then," said Potit, " did you not pay 15,000 rupees for the village, which you were going to get back by raising our rents ? "

" Why," said the merchant, " I did not for

a moment think of the villagers' suffering. It was merely an investment, as it might have been in sugar or piece-goods."

"Fool again," said Potit, "if you cannot see what must happen to us if your investment turns out well for you."

"I hope," said Jodu, "that now you have had a taste of our quality, you will let us alone for the future."

"I would as soon," said the merchant, "handle cobras and scorpions. But I will at least have the satisfaction of showing you up, and your police and lawyers, when I get back to Calcutta. Corruption and fraud shall at least be exposed, and perhaps punished. The High Court shall know the conduct which your magistrates and judges allow to pass unpunished."

"How do you propose," said the sub-inspector, "to accuse me of corruption?"

"I will do so in the public press," said the merchant.

"Well," said the sub-inspector, "many people do that, and then one of two things happens. Either my superiors take no notice, in which case it is not worth my while to pay any attention to what you say; or I am called on to defend my character, in which case it will be my painful duty to prosecute you or sue you for damages in the courts."

"And let me warn you," said Baykanta, "that I have no superiors to consider; but my character is my means of livelihood, and if I catch you saying a word against it to any one or anywhere, I shall at once institute legal proceedings."

"And so shall we all—we lawyers," said Koylesh.

"Remember," said Sital, "that you have disgusted by your stinginess and selfishness all those of your own side, so there is not a soul in the whole country-side who would give evidence for you."

"And who are you, to expose fraud and

corruption?" said Rám Ratan—"you who are as deep in them as any of us? You were ready enough to use them for the ruin of the other side."

"Oh! oh!" said Rámchandra to himself, "what shall I do? what shall I do, if I can't even speak my mind?"

And that is how Rámchandra the merchant went back to his shop in Calcutta, and did not become a great landlord.

To those who have no experience of rural India, the above account of the manner in which the law is made to serve local purposes may seem overdrawn. Those who have that experience, however, will probably consider it only a mild and imperfect picture of the reality. Where there is a Hurbullub on either side, the struggle is more equal, longer, and more complicated, as may be guessed. The Calcutta merchant was a stranger, unskilled in the ways of rural litigation, and made but a short stand.

A PEEP

INTO THE SONTHAL COUNTRY

A PEEP

INTO THE SONTHAL COUNTRY.

THE river Ganges, on its eastward course, after passing Patna and Bhaugulpore, and just before it finally enters the delta of Lower Bengal, has to pass by the northern foot of a range of low hills, which seem to push it off its course, for immediately on passing them it sweeps round to the south.

These low hills, which here and there reach a height of 2000 feet, extend southwards for about 100 miles, and form the nucleus of a revenue district whose area is 5500 square miles, and its population a million and three - quarters, known as the Sonthal Pergunnahs.

This, though only a single one of the forty-six districts of which the province of Lower Bengal is made up, contains much that deserves notice. There are the sacred Hindoo shrine of Baidyanath, one of the four holy places of Bengal; the ancient city of Rajmehal, once capital of the province; the old fort of Teliagurhi, once, in Mahomedan times, the key between Bengal and Behar, standing on a spur of the hill that plunges straight down into the river Ganges; and the Murgo Pass, through which, avoiding this stronghold, the Mahrattas poured into Bengal. There are everywhere remains of forts, and ancient traditions which tell of the days when chief fought with chief, and violence reigned. Again, there is an interesting mixture of races — the Hindoos from the plains of Bengal and Behar, and from the far-away west; the usual admixture of Mussulmans; the ancient aboriginal races of the Paharias, living on the hill-tops; the

Bhunyas, Khetowries, and Nats, holding the ground to the west of the hills; and above all, the Southals, whose name the district bears, and who have spread upwards from Orissa and the south, till now they are over-flowing even to the north of the Ganges.

The purpose of the following pages, how-ever, is to introduce the reader to a very in-teresting experiment in local self-government which was begun a hundred years ago, and entered on its present stage in 1856, the year before the Mutiny. It is an experi-ment in self - government of which the reader is not likely to hear much from the newspapers, which know little of these parts; or from the Congress, which has little to do with any but high - caste Hindoos; or from statistical returns or accounts, for it is not one of the various schemes of muni-cipal and district boards which have been established by statute. Yet, from a practi-cal point of view, it is more important, if less interesting, than all of these; just as field cultivation in England is more important than the rich man's hothouse or costly garden. It is a work for everyday use, carried on without much tending, dealing with a whole community; and, like most institutions that succeed, this one has its origin in the nature of the people and the country, rather than in the head of any law-maker, however wise or ingenious.

It would be tedious to attempt a complete description of the plant that has thus been nursed up in the Sonthal Pergunnahs. Perhaps, however, an idea of it can be given by showing the reader some of its roots in the past, and giving him a peep at it as it grows.

A hundred years ago, these hills were thickly covered with forest, full of game and beasts of prey. From time immemorial there lived a race of Paharias or highlanders, literally on the hill-tops, and not, like the

Highlanders of Scotland, in the glens below.
Like the old caterans, they would raid the
plains, and carry off their spoil of cattle and
other goods to their jungle fastnesses. Living
by hunting and plunder, they successfully
defied all the efforts of the Mahomedan
rulers of the land, and of their deputies the
great landlords, to subdue them. Accordingly,
when the British rule succeeded the Mahomedan, the Government withdrew this
country to the extent of some 1300 square
miles from the control of the landlords, took
it into its own hands, and deputed officers
to civilise and subdue the hillmen.

One of these officers, Cleveland by name,
who died at the early age of twenty-nine,
made himself a lasting name by his efforts
for the welfare of the hillmen. He found
that, far from being an incoherent mass of
lawless men, as their enemies described them,
the Paharias were strongly organised, each
hill - village under its Manjhi or headman,

and villages again in groups under Sardárs or chiefs, and Naibs (Nawab or Nabob is a form of this word) or deputies. He thought it possible to transform them from robbers and hunters perched on the hill-tops into peaceful cultivators of the plains below, and to this end he induced the Government to reserve the plains as a heritage rent-free for all Paharias who would come down, clear, and cultivate.

To enable them to live meantime, he got stipends for their chiefs, and he utilised the surplus energy of the young men by raising a regiment of paid and disciplined men, who, first as archers, and then as a regular regiment armed with muskets, were depended on to keep the peace in the hills. The regiment was known as the Hill Rangers.

Then he established, or rather strengthened and gave authority to, an assembly of hill chiefs, to whom, instead of to the ordinary courts, was given jurisdiction over all Paharias charged with offences against the law. Before this tribunal was held the great trial at Bhaugulpore, mentioned in a former chapter, of 120 men for a raid in which murders were committed and 900 head of cattle were lifted. Some were condemned to death, and many imprisoned.

What might have happened had Cleveland, with his great personal influence and the deep interest he took in the Paharias, lived to work out his own plans, we cannot tell. Possibly he might have got them down from their hills, to clear and occupy the plains and valleys below. As it is, they never came down, and there are their children still, perched upon the hill-tops, a feeble and waning race.

For over thirty years the straths and glens below and between the hills were left vacant for the hillmen to enter and take possession, but in vain. A change then came over the policy of the Government.

From the jungles of the south-west— Orissa and Chota Nagpore—there had come streaming into the country around swarms upon swarms of Sonthals. The Sonthals, unlike the Paharias, were very industrious, and great clearers of jungle. A Sonthal, when he sees a tree, says, "Here is a tree, let me cut it down," just as a Bengali says when he sees clean water, "Here is clean water, let me steep my jute in it." They were fond of hunting, skilful in the use of the bow, fond of all kinds of fun, and would dance, men and women, all through a moonlight night, to the music of the drum and the flute; accustomed to eat all kinds of food —grain, roots, mushrooms, leaves, fruits and nuts, animals and birds, and even snakes and ants; great drinkers of beer, which they made of rice; full of health and energy; mighty workers in earth, damming streams, and turning the sandy, boulder-strewn beds of torrents into terraced rice-fields; very

clannish among themselves, defying alike
the terrors of wild beasts and of malaria,
and turning uninhabited wastes into fields
of rice and maize. Such were the Sonthals.

They were very strongly organised among
themselves, being always grouped in villages,
each of which had its officials. The village
houses stood in a double row along the single
lane or *kulhi*, in which stood the *manjhi
than* or town-hall, an open shed supposed
to be inhabited by the village guardian
spirits — old man Manjhi and old woman
Manjhi.

Chief of the village officials was the Manjhi
or headman, generally the original clearer
or his representative, and he was the magistrate, controlled (for the Sonthals were very
republican) by a *panchayet* or council of
five elders. Then there came the Paramánik
or village field-bailiff, who looked after the
distribution of the fields ; the Naik or priest,
the Jog Manjhi or censor of morals, and the

Gorait or messenger. Among the duties of the Jog Manjhi is one both comical and useful. When all the rest get drunk, he is bound to remain sober and soothe quarrels. The influence of the 'Jog Manjhi to restore good feeling and prevent violence is said to be often magical, loud voices becoming mild and the raised hand sinking at a word from him.

The villages, again, were grouped under Purgunaits or district chiefs, who had their assistants, the Des Manjhi or chief constable, and Chakladar or revenue assistant. The Purgunait is the highest in rank among the Sonthals. But with them no single man or group of men is allowed to be master. Before any action is taken, there must always be a *durup* or session — the villagers meeting in session at the *manjhi than* or town-hall, in the *kulhi* or lane—and with them the *kulhi durup* is a ceremony of the first importance. For matters of more general importance the Purgunait calls a meeting of all his villages. Once a year all the Sonthals who can get together meet for their great annual hunt. This is in March, when the rice-harvest is over, before cultivation begins, when the *sal* forest is bare of leaves, and game can be most easily seen and followed. Here matters of general interest are discussed and settled, and here one Sonthal is as good as another, unpopular Purgunaits and Manjhies occasionally coming in for rough treatment.

To this people, so organised, the Government threw open the unoccupied valleys and glens of the Paharia reserve, this old-world Oklohama of fifty-five years ago, and there was a rush of adventurous spirits, who quickly, after their manner, formed villages, made clearings, and settled down, with their Purgunaits, Manjhies, and other officers, *kulhi durups*, great annual hunts, and all.

The Sonthals had — as who has not? —

X

their plagues and parasites, which followed and fastened on them. The Hindoo cultivators from the plains of Bengal and Behar, not bold enough themselves to clear jungle, would follow the Sonthal pioneers, and worry or cheat them out of the land they had reclaimed with so great labour. The money-lender and trader, out of a small debt, would run up against the Sonthal an ever-increasing score for compound interest, which, do what he would, seemed ever to be growing. The rent-collector was always trying to squeeze for his master and for himself more and more out of the hardy but simple toilers who had given the land its value. The myrmidons of the law — subordinates of the police and civil courts—were ever ready on their own account, or on account of any who would make it worth their while, to use upon the Sonthals the many means of irritation they possessed.

For protection from this class of men, the

Paharias had been withdrawn from the ordinary law and put under a special jurisdiction; but a like privilege was not extended to the Sonthals, who suffered accordingly. Their villages were taken out of the control of their Manjhics, and handed over to speculators; money-lenders settled down in every village, fastened on ryot after ryot, and sucked their blood; the subordinates of the police and the law courts oppressed and extorted greedily; and within twenty years of their admission into the reserve the Sonthals, law-abiding and good-natured as they usually were, became desperate, and the more spirited formed themselves into robber-gangs.

Things move leisurely in the public offices. Steps for the removal of the acknowledged grievances of the Sonthals were still under discussion, and even had made some progress towards being actually put in practice, when the Sonthals took the affair into their own hands and rose. Kanoo and Sidoo, two brothers who claimed to be inspired from heaven, started a *hool*. The *Hool* or national rising, like the Jehad of Islam and the Crusade of the middle ages, was a race movement, gaining much of its force from other causes, but organised under the banner of religion.

The Sonthal *hool* was an uprising against the oppression of the Dikku, and an attempt to throw off his yoke. To the Sonthal there are three main divisions of men, besides the Paharias—the Sonthal or "man" proper, the Sahib or European, and the Dikku. The Dikkus are the plainsmen of Bengal and Behar, nearly all Hindoos. With the Sahibs the Sonthals have always been on the best of terms; but they hated, and still hate, the Dikku, looking on him as a cunning and treacherous enemy, ever on the watch for a chance of outwitting or cheating the simple Sonthal. The Dikku personifies all the plagues,—the ousting cultivator, the usurer, the policeman, the process-server, and the rent-collector—parasites all. The Hindoo, on his side, despises the Sonthal as being of no caste, and looks on him—like the bees and the oxen of the Roman poet, toiling, not for themselves—as one to be smoked out of the fields he has made, or to bear meekly the oppressive yoke of the usurer's bond. The Sonthals rose, not against the English, but to free themselves from the meshes of petty tyranny with which the Dikkus or plainsmen were strangling them.

The first blow was struck at a place called Pachkattia. Mohesh Darogah, combining in himself the offices of policeman and rent-collector, was a man hateful to the Sonthals above even the usurers. Hearing one day from a usurer that the Sonthals had met, and were talking over their grievances, he went with his informant, his Burkandazes (the nearest European equivalent for these worthies is the Spanish *alguazil*), and a cartload of

ropes, to the place of meeting at Pachkattia.
He asked why they were gathered there.
"Never mind," said they. "Disperse at
once," said he, "or I will tie you all up."
"You had better tie us up, for if you do not,
we will kill you," said the Sonthals, and high
words passed on both sides. "You a Darogah!" (police inspector) said they. "You
extort and oppress. You are no Darogah,
but a dacoit." And they ended by falling
on the Darogah and the usurer with their
fursas (battle-axes) and killing them. The
heather was on fire, and the flame of the *hool*
spread.

Grim were some of the scenes, of which a
glimpse is caught through the mist of confusion which hid the struggle. A European
went out on an elephant, and his two sons
on horseback, armed with guns, to fight a party
of Sonthals. All three were shot down with
arrows. A European family at Sahibgunj, on
the Ganges, took boat as the Sonthals ap-

proached, and, keeping out of reach of the arrows, killed many with their firearms. A native landlord was besieged by the Sonthals in his house, which was set on fire. He took refuge in a tank, where he was killed with arrows, and his body chopped up on a stone into twenty-two pieces, "one for each of his ancestors." The stone on which his body was chopped up still stands at his house door, and his two widows, who took refuge in the jungle, still live, one in the old house, and the other close by.

As a grim joke, the Sonthals took a usurer who had made himself obnoxious, and chopped off his legs at the knees, crying " Four annas." (The whole rupee is sixteen annas, and this meant that a fourth had been paid.) Then they cut his legs off at the thighs, crying "Eight annas." Then his arms, crying "Twelve annas ; " and last they cut off his head, crying " *Farkati !* " (quittance in full).

The impression of this savage joke still

remains; for a year or two ago, when a mad
Sonthal killed a usurer on the east side of
the district, the usurers began pouring out
of it at the west side, forty miles away, crying that there was another *hool*.

The regiment of Hill Rangers, which had
sufficed to keep order among the Paharias,
proved too weak to hold down the Sonthals,
and was disbanded a year or two later. I
saw about forty veterans, pensioners of
this regiment, go through their drill after
twenty - five years' disuse. The men were
nearly all old, many decrepit; they had
but one cap and two tunics among them,
and for muskets they had bamboo staves.
But they marched with precision, and went
through their movements with the machine-
like rigidity of the old style. It was a curious glimpse of bygone days, about to vanish
for ever, in that little valley among the hills.

After much outrage and bloodshed, and
much suffering on both sides, the rising

was, with the help of troops brought in from without, at length put down. These very troops joined next year in the great Mutiny, and had themselves in turn to be crushed.

After the *hool* had come to an end, the Government, recognising that it had been brought about mainly through the oppressions exercised by Government servants, withdrew from the control of the ordinary courts of law and system of police the part of the country known ever since as the Sonthal Pergunnahs, and established for this district a system of its own, founded to a great extent on the indigenous organisation of the people, which was utilised to the utmost, under European supervision, for the government of the people by themselves. Every village had its head, who was magistrate and police and revenue officer in one. The usurers and traders, instead of being allowed to spread about in agricultural villages, were, especially in the Government reserve, gathered into "bazars" or trading villages of their own, each of which had also its Choudhry or mayor, who was manager of all the village business, such as police, sanitation, and so on, and represented the village in its dealings with Government officers.

It is one of the most important duties of Government officers in the Sonthal Pergunnahs, almost alone among the districts of Bengal, to look after the headmen of the eleven thousand village areas into which it is divided, taking steps to weed out men who are unfit, or have been guilty of misconduct, and to replace the waste caused by death or dismissal of headmen. Every time a headman is dismissed, or dies without heirs, his place is supplied by election, and it is an election scene that the reader is now invited to witness.

I had, in the course of my annual tour, reached the village of Barhait, a bazar or trading village containing about 200 houses, and situated some fourteen miles from the railway. It is a great trading centre, and the centre of a Purgunait's circle. The Choudhry had been dismissed for misconduct some years before, and we had never been able to find a fit successor. A proposal had indeed been made to appoint the Purgunait, Boloram, to the post; but this arrangement was considered unsuitable, as he was chief of the Sonthals in the outside villages, and it was thought desirable to have a different man to be chief of the Dikkus of the bazar.

I thought this would be a good opportunity to fix on the new Choudhry, and accordingly called a meeting of the bazar people at the bungalow.

The bungalow is a shelter of mud walls and a roof put up and repaired at the Purgunait's centre of each circle, to enable Government officers to visit it at any time of the year.

Picture then the Sahib, in an old suit of blue serge, seated on a chair in the shady verandah of the bungalow, with Boloram Purgunait standing by his side, and on the ground outside the two candidates—a Bengali trader and the Momin (a sort of Mahomedan) Poramanik of a neighbouring village—with a crowd of about 300 people behind them. Barhait is always a great place for crowds. The sightseers are sifted from the voters and made to stand back, and the group of about seventy ratepayers who have a voice in the election are bidden to seat themselves in front of the Sahib. They are made to sit down,—first, that the air may get in; then, because standing men have a tendency to edge forward and prevent all but a few from seeing and hearing; and lastly, because natives can think better when squatting on their heels. So we have the Sahib in the verandah, the Purgunait beside him; below, on the ground, the two candidates standing, and the group of voters seated;

and in the background a ring of men, women, and children looking on. Till the sun goes down and the shadows lengthen there will not be room for these last in the shade of the bungalow; so for a time yet the sun lights up their swarthy skin, white clothing, and glancing eyes.

It is then explained by the Sahib that the meeting is called to choose a Choudhry for the village, and that the two candidates who have come forward, the Bengali and the Momin, are before the voters. All who vote for the Bengali are now told to stand. No one moves. Then all who vote for the Momin: still all remain seated. Thinking they may not understand, the Sahib sends the Bengali to one side, and the Momin to the other, and tells all who favour either candidate to go beside him. No one budges—all stolid and expressionless. Determined to get a vote, the Sahib sets the Purgunait in the middle, and says that all on

one side will count for the Momin, and all on
the other for the Bengali. One or two get up
and cross over to the side of the Bengali, but
very few have moved. The Momin, as the
candidate less favoured, is now called on for
a speech, and he delivers a vigorous harangue,
denouncing the Bengali as a bad character
and receiver of stolen property (which there
is reason to suspect he is).

Plainly perceiving that the interest of the
voters for either candidate is of the coldest,
the Sahib asks if there are no respectable
traders in the bazar who would be more ac-
ceptable. The names of three men are given,
all of whom, however, have been asked to
take office, and have refused. Things are at
a dead-lock.

At last up gets one of the voters and says,
" The fact is, Sahib, there is only one man
who will do for us — Boloram Purgunait.
His father Shám was Purgunait and Choudhry both, and he is the only man who

can keep the peace between us traders and
the Sonthals." Now Boloram has not been
admitted as a candidate, as he, being chief of
the Sonthals, was considered an undesirable
man to have as Choudhry of a Dikku bazar.
But in this country custom is powerful, so
the Sahib says, "If such was your old
custom, I am unwilling to break old customs. I will agree to your having Boloram
as Choudhry."

What a change! The dull faces light up;
the people jump to their feet and press in,
their eyes glistening, some with tears; and
they all cry, "Now all will be well! we have
got the right man!" The two candidates
withdraw their claims in favour of the favourite, who gracefully yields to the popular
wish. And so the business is settled. The
Choudhry has been appointed, and, until he
resigns, or breaks the law, or is found to have
oppressed the villagers and lost their confidence, he will hold his office. Should any

of these things happen, there will be another election.

Is it possible to convey to the mind of the English reader a sense of the meaning and importance of this short and simple proceeding, imperfectly understood as it is by most people even in India? Perhaps it is the sort of thing that will be more readily appreciated in free England.

The villagers do not get a perfect chief, but at all events they get the best man they can produce. He represents the village to the State and to the outside world — has charge and keeps the key, as it were. For this privilege of privacy under a guardian of their own choice, the only condition required by Government of the villagers is that that guardian shall not be a man who, from his past conduct or character, is likely to thwart the Government or the law—by harbouring thieves or concealing crime, for instance.

The knowledge that he holds his office by

virtue of the confidence placed in him by the State and by the villagers, and may lose the office if he forfeits confidence, is as strong a working check as can be devised on the tendency of all Orientals vested with power to use that power for their own advantage.

The presence of this village head in each village has enabled the Government to dispense with, or at all events reduce and keep within bounds, that swarm of parasites which follow its rule elsewhere, and are the despair of all zealous rulers—the petty police, process-servers, rent-collectors, and petty lawyers. Had we to deal individually with each of the one and three-quarter millions of people, instead of with the eleven thousand heads of villages, there would be a great many more of these. They are there — in such a soil how can they but grow? — but the authority of the village head takes up much of the ground, and so they can be kept within bounds.

Y

Then, as has been seen, the system is practicable on a large scale, because much of the time and labour of officials is not needed to keep it going. This particular election, the work of a couple of hours in an afternoon, was held over five years ago, and there has been no occasion since that time to revise the choice then made. Such a revision is only necessary on the death or resignation or proved misconduct of the holder of the office.

Looked at from the point of view of the villagers, the choice of their chief is of the utmost importance. I shall never forget the glisten of the eyes that advanced on me when the right man was at last named. There was no doubt that these men were in earnest, and believed that on this choice depended the happiness or misery of a certain number of households.

Some may say, " Why, then, give such power to any man? Why not keep it for the State?" But it is not in the power of the rulers of great provinces, or even districts, to prevent oppression or give good government, otherwise than by putting power in the hands of the right men, and keeping it out of the hands of the wrong.

In a village where no chief has been recognised by Government, the privacy it gains by having such a chief is lost, but there is still a master. The master is a fact, whether recognised by the State or not. The question in the Oriental village is not, "Master or no master?" but "Who is master?" In villages where no man is recognised by Government, it is generally a resident proprietor, an unchecked despot; or an absentee proprietor acting through an ill-paid servant, intent on feathering his nest; or the local usurer, with his hook in the nose of every ryot. If the villagers are independent enough to defy all these powers, there springs up from time to time that institution so familiar in Bengal, the *daládali* or faction feud, which differs from war only in that it is on a smaller scale, and the injuries done in it are less open. The landlord, the rent-collector, the usurer, the priest, the petty lawyer, the village ruffian,—all these furnish materials for a despotism if they agree; for a severe struggle if they quarrel. There must be a master.

Herein the election of the kind we have now to do with differs from the English election. The English election is the process by which free men determine who is to execute their common will; while the election in the Indian village is held to find out who is master, and whether he is a fit person to be intrusted by Government with its authority. The master is a fact that existed before the election, and would exist if there were no election.

If the above description was faithful, it will appear that the answer to the question, " Who is master ? " did not lie on the surface. The villagers might easily have elected a man they were afraid of, or cared nothing for, instead of the man they liked best.

The question whether the man recognised is fit to hold authority from Government, is one that should obviously be left to be answered by a confidential officer of Government. The standard of excellence need not be high, for the alternative is a sorry one; but obviously the man should at least be well disposed to the State and the law.

So long, however, as he is possible from a Government point of view, the recognition of him by the Government will do good by the restraint of his responsibility, and by the lesson taught the villagers that there is a way of escape. Natives of Bengal are victims of oppression, chiefly because of their own nature, and to educate men out of their own nature is at best but a slow process. This is not the last, but an early step in the process.

The difference between having the right man and having the wrong man at the head of a village is great. With the former, things go smoothly and well; with the latter, there is constant friction. In the Sonthal Pergunnahs, politics consist for the most part in adjusting and keeping in order this important part of the social mechanism.

The chief significance, however, of the Sonthal system is that, as its name implies, it is the result of education—not of a subject people by a Government, but of a Government by a subject people. The free instincts and straightforward disposition of the Sonthals could not endure the plagues which the long-suffering patience of their neighbours the Bengalis had caused to be regarded as natural and unavoidable, and they forced the Government to devise a more reasonable way of management.

This way has hardly as yet been extended outside the one district, and, even within it, has never been fully developed. By far the greater part of the officials of the province, with the full sympathy of the great and influential army of clerks, police, lawyers, and usurers, look forward to the coming of the good time when it will be possible to let loose on the Sonthals all the parasites of so-called civilisation which it is the privilege of their neighbours to endure. I hope that time will never come. Rather do I look forward to the time when this excellent system shall be extended to the whole province. Why not? It is not peculiar to the Sonthals, but was once common,—as may it become again!—to all races in the country. This is my earnest hope, because I know no other method of laying a firm foundation for a strong and free fabric of local self-government. I will not venture to predict what shape such a fabric would take. An architect may do this of the lifeless building he proposes to construct. Of a living organism, however, we can say no more than that, if it is healthy and free, it is likely to adapt itself to the needs of the community and work well. No stronger proof of the spirit it is of can be found than the fact that a community of Hindoos and Mussulmans chose as their chief a man of the despised Sonthal race. Those who know them know what that means.

THE END.

PRINTED BY WILLIAM BLACKWOOD AND SONS.

Catalogue

of

Messrs Blackwood & Sons'

Publications

PHILOSOPHICAL CLASSICS FOR ENGLISH READERS.

EDITED BY WILLIAM KNIGHT, LL.D.,
Professor of Moral Philosophy in the University of St Andrews.

In crown 8vo Volumes, with Portraits, price 3s. 6d.

Contents of the Series.

DESCARTES, by Professor Mahaffy, Dublin.—BUTLER, by Rev. W. Lucas Collins, M.A.—BERKELEY, by Professor Campbell Fraser.—FICHTE, by Professor Adamson, Owens College, Manchester. — KANT, by Professor Wallace, Oxford.—HAMILTON, by Professor Veitch, Glasgow. — HEGEL, by the Master of Balliol. — LEIBNIZ, by J. Theodore Merz.—VICO, by Professor Flint, Edinburgh.—HOBBES, by Professor Croom Robertson.—HUME, by the Editor.—SPINOZA, by the Very Rev. Principal Caird, Glasgow. — BACON: Part 1. The Life, by Professor Nichol.—BACON: Part II. Philosophy, by the same Author.—LOCKE, by Professor Campbell Fraser.

FOREIGN CLASSICS FOR ENGLISH READERS.

EDITED BY MRS OLIPHANT.

In crown 8vo, 2s. 6d.

Contents of the Series.

DANTE, by the Editor.—VOLTAIRE, by General Sir E. B. Hamley, K.C.B.—PASCAL, by Principal Tulloch.—PETRARCH, by Henry Reeve, C.B.—GOETHE, by A. Hayward, Q.C.—MOLIÈRE, by the Editor and F. Tarver, M.A.—MONTAIGNE, by Rev. W. L. Collins, M.A.—RABELAIS, by Walter Besant, M.A.—CALDERON, by E. J. Hasell. — SAINT SIMON, by Clifton W. Collins, M.A. — CERVANTES, by the Editor. — CORNEILLE AND RACINE, by Henry M. Trollope. — MADAME DE SÉVIGNÉ, by Miss Thackeray.—LA FONTAINE, AND OTHER FRENCH FABULISTS, by Rev. W. Lucas Collins, M.A.—SCHILLER, by James Sime, M.A., Author of 'Lessing, his Life and Writings.'—TASSO, by E. J. Hasell. — ROUSSEAU, by Henry Grey Graham.—ALFRED DE MUSSET, by C. F. Oliphant.

ANCIENT CLASSICS FOR ENGLISH READERS.

EDITED BY THE REV. W. LUCAS COLLINS, M.A.

Complete in 28 Vols. crown 8vo, cloth, price 2s. 6d. each. And may also be had in 14 Volumes, strongly and neatly bound, with calf or vellum back, £3, 10s.

Contents of the Series.

HOMER: THE ILIAD, by the Editor.—HOMER: THE ODYSSEY, by the Editor.—HERODOTUS, by George C. Swayne, M.A.—XENOPHON, by Sir Alexander Grant, Bart., LL.D. — EURIPIDES, by W. B. Donne.—ARISTOPHANES, by the Editor.—PLATO, by Clifton W. Collins, M.A.—LUCIAN, by the Editor. — ÆSCHYLUS, by the Right Rev. the Bishop of Colombo. — SOPHOCLES, by Clifton W. Collins, M.A. — HESIOD AND THEOGNIS, by the Rev. J. Davies, M.A.—GREEK ANTHOLOGY, by Lord Neaves. — VIRGIL, by the Editor.—HORACE, by Sir Theodore Martin, K.C.B.—JUVENAL, by Edward Walford, M.A. — PLAUTUS AND TERENCE, by the Editor—THE COMMENTARIES OF CÆSAR, by Anthony Trollope.—TACITUS, by W. B. Donne.—CICERO, by the Editor.—PLINY'S LETTERS, by the Rev. Alfred Church, M.A., and the Rev. W. J. Brodribb, M.A. — LIVY, by the Editor.—OVID, by the Rev. A. Church, M.A.—CATULLUS, TIBULLUS, AND PROPERTIUS, by the Rev. Jas. Davies, M.A. — DEMOSTHENES, by the Rev. W. J. Brodribb, M.A.—ARISTOTLE, by Sir Alexander Grant, Bart., LL.D.—THUCYDIDES, by the Rev. A. Church, M.A.—LUCRETIUS, by W. H. Mallock, M.A.—PINDAR, by the Rev. F. D. Morice, M.A.

Saturday Review.—"It is difficult to estimate too highly the value of such a series as this in giving 'English readers' an insight, exact as far as it goes, into those olden times which are so remote, and yet to many of us so close."

CATALOGUE

OF

MESSRS BLACKWOOD & SONS'

PUBLICATIONS.

———•———

ALISON.
History of Europe. By Sir ARCHIBALD ALISON, Bart., D.C.L.
1. From the Commencement of the French Revolution to
the Battle of Waterloo.
LIBRARY EDITION, 14 vols., with Portraits. Demy 8vo, £10, 10s.
ANOTHER EDITION, in 20 vols. crown 8vo, £6.
PEOPLE'S EDITION, 13 vols. crown 8vo, £2, 11s.

2. Continuation to the Accession of Louis Napoleon.
LIBRARY EDITION, 8 vols. 8vo, £6, 7s. 6d.
PEOPLE'S EDITION, 8 vols. crown 8vo, 34s.

Epitome of Alison's History of Europe. Thirtieth Thousand, 7s. 6d.

Atlas to Alison's History of Europe. By A. Keith Johnston.
LIBRARY EDITION, demy 4to, £3, 3s.
PEOPLE'S EDITION, 31s. 6d.

Life of John Duke of Marlborough. With some Account of
his Contemporaries, and of the War of the Succession. Third Edition. 2 vols.
8vo. Portraits and Maps, 30s.

Essays : Historical, Political, and Miscellaneous. 3 vols.
demy 8vo, 45s.

ACROSS FRANCE IN A CARAVAN : BEING SOME ACCOUNT
OF A JOURNEY FROM BORDEAUX TO GENOA IN THE "ESCARGOT," taken in the Winter
1889-90. By the Author of 'A Day of my Life at Eton.' With fifty Illustrations
by John Wallace, after Sketches by the Author, and a Map. Cheap Edition,
demy 8vo, 7s. 6d.

ACTA SANCTORUM HIBERNIÆ ; Ex Codice Salmanticensi.
Nunc primum integre edita opera CAROLI DE SMEDT et JOSEPHI DE BACKER, e
Soc. Jesu, Hagiographorum Bollandianorum ; Auctore et Sumptus Largiente
JOANNE PATRICIO MARCHIONE BOTHAE. In One handsome 4to Volume, bound in
half roxburghe, £2, 2s.; in paper cover, 31s. 6d.

AGRICULTURAL HOLDINGS ACT, 1883. With Notes by a
MEMBER OF THE HIGHLAND AND AGRICULTURAL SOCIETY. 8vo, 3s. 6d.

AIKMAN.
Manures and the Principles of Manuring. By C. M. AIKMAN,
D.Sc., F.R.S.E., &c., Professor of Chemistry, Glasgow Veterinary College;
Examiner in Chemistry, University of Glasgow, &c. Crown 8vo, 6s. 6d.

Farmyard Manure : Its Nature, Composition, and Treatment.
Crown 8vo 1s. 6d.

AIRD.   Poetical Works of Thomas Aird.   Fifth Edition, with
Memoir of the Author by the Rev. JARDINE WALLACE, and Portrait.   Crown 8vo,
7s. 6d.

ALLARDYCE.
Balmoral :   A Romance of the Queen's Country.   By ALEX-
ANDER ALLARDYCE.   3 vols. crown 8vo, 25s. 6d.
Earlscourt : A Novel of Provincial Life.   3 vols. crown 8vo,
25s. 6d.
The City of Sunshine.   New and Revised Edition.   Crown
8vo, 6s.
Memoir of the Honourable George Keith Elphinstone, K.B.,
Viscount Keith of Stonehaven, Marischal, Admiral of the Red.   8vo, with Por-
trait, Illustrations, and Maps, 21s.

ALMOND.   Sermons by a Lay Head-master.   By HELY HUTCH-
INSON ALMOND, M.A. Oxon., Head-master of Loretto School.   Crown 8vo, 5s.

ANCIENT CLASSICS FOR ENGLISH READERS.   Edited
by Rev. W. LUCAS COLLINS, M.A.   Price 2s. 6d. each.   *For List of Vols., see p. 2.*

ANNALS OF A FISHING VILLAGE.   By "A SON OF THE
MARSHES."   *See page 28.*

AYTOUN.
Lays of the Scottish Cavaliers, and other Poems.   By W.
EDMONDSTOUNE AYTOUN, D.C.L., Professor of Rhetoric and Belles-Lettres in the
University of Edinburgh.   New Edition.   Fcap. 8vo, 3s. 6d.
ANOTHER EDITION.   Fcap. 8vo, 7s. 6d.
CHEAP EDITION.   1s.   Cloth, 1s. 3d.
An Illustrated Edition of the Lays of the Scottish Cavaliers.
From designs by Sir NOEL PATON.   Small 4to, in gilt cloth, 21s.
Bothwell : a Poem.   Third Edition.   Fcap., 7s. 6d.
Poems and Ballads of Goethe.   Translated by Professor
AYTOUN and Sir THEODORE MARTIN, K.C.B.   Third Edition.   Fcap., 6s.
Bon Gaultier's Book of Ballads.   By the SAME.   Fifteenth
Edition.   With Illustrations by Doyle, Leech, and Crowquill.   Fcap. 8vo, 5s.
The Ballads of Scotland.   Edited by Professor AYTOUN.
Fourth Edition.   2 vols. fcap. 8vo, 12s.
Memoir of William E. Aytoun, D.C.L.   By Sir THEODORE
MARTIN, K.C.B.   With Portrait.   Post 8vo, 12s.

BACH.
On Musical Education and Vocal Culture.   By ALBERT B.
BACH.   Fourth Edition.   8vo, 7s. 6d.
The Principles of Singing.   A Practical Guide for Vocalists
and Teachers.   With Course of Vocal Exercises.   Crown 8vo, 6s.
The Art of Singing.   With Musical Exercises for Young
People.   Crown 8vo, 3s.
The Art Ballad : Loewe and Schubert.   With Musical Illus-
trations.   With a Portrait of LOEWE.   Third Edition.   Small 4to, 5s.

BAIRD LECTURES.
Theism.   By Rev. Professor FLINT, D.D., Edinburgh.   Eighth
Edition.   Crown 8vo, 7s. 6d.
Anti-Theistic Theories.   By the SAME.   Fifth Edition.   Crown
8vo, 10s. 6d.
The Inspiration of the Holy Scriptures.   By Rev. ROBERT
JAMIESON, D.D.   Crown 8vo, 7s. 6d.

BAIRD LECTURES.
The Early Religion of Israel. As set forth by Biblical Writers
and modern Critical Historians. By Rev. Professor ROBERTSON, D.D., Glasgow.
Fourth Edition. Crown 8vo, 10s. 6d.
The Mysteries of Christianity. By Rev. Professor CRAWFORD,
D.D. Crown 8vo, 7s. 6d.
Endowed Territorial Work : Its Supreme Importance to the
Church and Country. By Rev. WILLIAM SMITH, D.D. Crown 8vo, 6s.
BALLADS AND POEMS. By MEMBERS OF THE GLASGOW
BALLAD CLUB. Crown 8vo, 7s. 6d.
BANNATYNE. Handbook of Republican Institutions in the
United States of America. Based upon Federal and State Laws, and other reli-
able sources of information. By DUGALD J. BANNATYNE, Scotch Solicitor, New
York; Member of the Faculty of Procurators, Glasgow. Crown 8vo, 7s. 6d.
BELLAIRS.
The Transvaal War, 1880-81. Edited by Lady BELLAIRS.
With a Frontispiece and Map. 8vo, 15s.
Gossips with Girls and Maidens, Betrothed and Free. New
Edition. Crown 8vo, 3s. 6d. Cloth, extra gilt edges, 5s.
BELLESHEIM. History of the Catholic Church of Scotland.
From the Introduction of Christianity to the Present Day. By ALPHONS BEL-
LESHEIM, D.D., Canon of Aix-la-Chapelle. Translated, with Notes and Additions,
by D. OSWALD HUNTER BLAIR, O.S.B., Monk of Fort Augustus. Complete in
4 vols. demy 8vo, with Maps. Price 12s. 6d. each.
BENTINCK. Racing Life of Lord George Cavendish Bentinck,
M.P., and other Reminiscences. By JOHN KENT, Private Trainer to the Good-
wood Stable. Edited by the Hon. FRANCIS LAWLEY. With Twenty-three full-
page Plates, and Facsimile Letter. Third Edition. Demy 8vo, 25s.
BESANT.
The Revolt of Man. By WALTER BESANT. Tenth Edition.
Crown 8vo, 3s. 6d.
Readings in Rabelais. Crown 8vo, 7s. 6d.
BEVERIDGE.
Culross and Tulliallan ; or Perthshire on Forth. Its History
and Antiquities. With Elucidations of Scottish Life and Character from the
Burgh and Kirk-Session Records of that District. By DAVID BEVERIDGE. 2 vols.
8vo, with Illustrations, 42s.
Between the Ochils and the Forth ; or, From Stirling Bridge
to Aberdour. Crown 8vo, 6s.
BICKERDYKE. A Banished Beauty. By JOHN BICKERDYKE,
Author of ' Days in Thule, with Rod, Gun, and Camera,' ' The Book of the All-
Round Angler,' ' Curiosities of Ale and Beer,' &c. With Illustrations. Crown
8vo, 6s.
BIRCH.
Examples of Stables, Hunting-Boxes, Kennels, Racing Estab-
lishments, &c. By JOHN BIRCH, Architect, Author of ' Country Architecture,'
&c. With 30 Plates. Royal 8vo, 7s.
Examples of Labourers' Cottages, &c. With Plans for Im-
proving the Dwellings of the Poor in Large Towns. With 34 Plates. Royal 8vo,
7s.
Picturesque Lodges. A Series of Designs for Gate Lodges,
Park Entrances, Keepers', Gardeners', Bailiffs', Grooms', Upper and Under Ser-
vants' Lodges, and other Rural Residences. With 16 Plates. 4to, 12s. 6d.
BLACK. Heligoland and the Islands of the North Sea. By
WILLIAM GEORGE BLACK. Crown 8vo, 4s.
BLACKIE.
Lays and Legends of Ancient Greece. By JOHN STUART
BLACKIE, Emeritus Professor of Greek in the University of Edinburgh. Second
Edition. Fcap. 8vo, 5s.

BLACKIE.

The Wisdom of Goethe. Fcap. 8vo. Cloth, extra gilt, 6s.

Scottish Song : Its Wealth, Wisdom, and Social Significance.
Crown 8vo. With Music. 7s. 6d.

A Song of Heroes. Crown 8vo, 6s.

BLACKMORE. The Maid of Sker. By R. D. BLACKMORE,
Author of 'Lorna Doone,' &c. New Edition. Crown 8vo, 6s.

BLACKWOOD.

Blackwood's Magazine, from Commencement in 1817 to October 1894. Nos. 1 to 948, forming 155 Volumes.

Index to Blackwood's Magazine. Vols. 1 to 50. 8vo, 15s.

Tales from Blackwood. First Series. Price One Shilling each,
in Paper Cover. Sold separately at all Railway Bookstalls.
They may also be had bound in 12 vols., cloth, 18s. Half calf, richly gilt, 30s.
Or the 12 vols. in 6, roxburghe, 21s. Half red morocco, 28s.

Tales from Blackwood. Second Series. Complete in Twenty-
four Shilling Parts. Handsomely bound in 12 vols., cloth, 30s. In leather back,
roxburghe style, 37s. 6d. Half calf, gilt, 52s. 6d. Half morocco, 55s.

Tales from Blackwood. Third Series. Complete in Twelve
Shilling Parts. Handsomely bound in 6 vols., cloth, 15s.; and in 12 vols., cloth,
18s. The 6 vols. in roxburghe, 21s. Half calf, 25s. Half morocco, 28s.

Travel, Adventure, and Sport. From 'Blackwood's Magazine.'
Uniform with 'Tales from Blackwood.' In Twelve Parts, each price 1s. Handsomely bound in 6 vols., cloth, 15s. And in half calf, 25s.

New Educational Series. *See separate Catalogue.*

New Uniform Series of Novels (Copyright).
Crown 8vo, cloth. Price 3s. 6d. each. Now ready :—

THE STORY OF MARGRÉDEL. By D. Storrar Meldrum
MISS MARJORIBANKS. By Mrs Oliphant
THE PERPETUAL CURATE, and THE RECTOR. By the Same.
SALEM CHAPEL, and THE DOCTOR'S FAMILY. By the Same
A SENSITIVE PLANT. By E. D. Gerard.
LADY LEE'S WIDOWHOOD. By General Sir E. B. Hamley.
KATIE STEWART, and other Stories. By Mrs Oliphant.
VALENTINE, AND HIS BROTHER. By the Same.
SONS AND DAUGHTERS. By the Same.
MARMORNE. By P. G. Hamerton.
REATA. By E. D. Gerard.
BEGGAR MY NEIGHBOUR. By the Same.

THE WATERS OF HERCULES. By the Same.
FAIR TO SEE. By L. W. M. Lockhart.
MINE IS THINE. By the Same.
DOUBLES AND QUITS. By the Same.
HURRISH. By the Hon. Emily Lawless.
ALTIORA PETO. By Laurence Oliphant.
PICCADILLY. By the Same. With Illustra-
tions.
THE REVOLT OF MAN. By Walter Besant.
LADY BABY. By D. Gerard.
THE BLACKSMITH OF VOE. By Paul Cushing.
THE DILEMMA. By the Author of 'The
Battle of Dorking.'
MY TRIVIAL LIFE AND MISFORTUNE. By A
Plain Woman.
POOR NELLIE. By the Same.

Others in preparation.

Standard Novels. Uniform in size and binding. Each
complete in one Volume.

FLORIN SERIES, Illustrated Boards. Bound in Cloth, 2s. 6d.

TOM CRINGLE'S LOG. By Michael Scott.
THE CRUISE OF THE MIDGE. By the Same.
CYRIL THORNTON. By Captain Hamilton.
ANNALS OF THE PARISH. By John Galt.
THE PROVOST, &c. By the Same.
SIR ANDREW WYLIE. By the Same.
THE ENTAIL. By the Same.
MISS MOLLY. By Beatrice May Butt.
REGINALD DALTON. By J. G. Lockhart.

PEN OWEN. By Dean Hook.
ADAM BLAIR. By J. G. Lockhart.
LADY LEE'S WIDOWHOOD. By General Sir E.
B. Hamley.
SALEM CHAPEL. By Mrs Oliphant.
THE PERPETUAL CURATE. By the Same.
MISS MARJORIBANKS. By the Same.
JOHN : A Love Story. By the Same.

BLACKWOOD.
Standard Novels.
SHILLING SERIES, Illustrated Cover. Bound in Cloth, 1s. 6d.

THE RECTOR, and THE DOCTOR'S FAMILY. SIR FRIZZLE PUMPKIN, NIGHTS AT MESS,
By Mrs Oliphant. &c.
THE LIFE OF MANSIE WAUCH. By D. M. THE SUBALTERN.
Moir. LIFE IN THE FAR WEST. By G. F. Ruxton.
PENINSULAR SCENES AND SKETCHES. By VALERIUS: A Roman Story. By J. G.
F. Hardman. Lockhart.

BON GAULTIER'S BOOK OF BALLADS. Fifteenth Edition. With Illustrations by Doyle, Leech, and Crowquill. Fcap. 8vo, 5s.

BONNAR. Biographical Sketch of George Meikle Kemp, Architect of the Scott Monument, Edinburgh. By THOMAS BONNAR, F.S.A. Scot., Author of 'The Present Art Revival,' &c. With Three Portraits and numerous Illustrations. Post 8vo, 7s. 6d.

BRADDON. Thirty Years of Shikar. By Sir EDWARD BRADDON, K.C.M.G. With numerous Illustrations. In 1 vol. demy 8vo. [*In the press.*

BROUGHAM. Memoirs of the Life and Times of Henry Lord Brougham. Written by HIMSELF. 3 vols. 8vo, £2, 8s. The Volumes are sold separately, price 16s. each.

BROWN. The Forester: A Practical Treatise on the Planting and Tending of Forest-trees and the General Management of Woodlands. By JAMES BROWN, LL.D. Sixth Edition, Enlarged. Edited by JOHN NISBET, D.Œc., Author of 'British Forest Trees,' &c. In 2 vols. royal 8vo, with 350 Illustrations, 42s. net.

BROWN. Stray Sport. By J. MORAY BROWN, Author of 'Shikar Sketches,' 'Powder, Spur, and Spear,' 'The Days when we went Hog-Hunting.' 2 vols. post 8vo, with Fifty Illustrations, 21s.

BROWN. A Manual of Botany, Anatomical and Physiological. For the Use of Students. By ROBERT BROWN, M.A., Ph.D. Crown 8vo, with numerous Illustrations, 12s. 6d.

BROWN. The Book of the Landed Estate. Containing Directions for the Management and Development of the Resources of Landed Property. By ROBERT E. BROWN, Factor and Estate Agent. Royal 8vo, with Illustrations, 21s.

BRUCE.
In Clover and Heather. Poems by WALLACE BRUCE. New and Enlarged Edition. Crown 8vo, 4s. 6d.
A limited number of Copies of the First Edition, on large hand-made paper, 12s. 6d.

Here's a Hand. Addresses and Poems. Crown 8vo, 5s.
Large Paper Edition, limited to 100 copies, price 21s.

BRYDALL. Art in Scotland; its Origin and Progress. By ROBERT BRYDALL, Master of St George's Art School of Glasgow. 8vo, 12s. 6d.

BUCHAN. Introductory Text-Book of Meteorology. By ALEXANDER BUCHAN, LL.D., F.R.S.E., Secretary of the Scottish Meteorological Society, &c. New Edition. Crown 8vo, with Coloured Charts and Engravings.
[*In preparation.*

BUCHANAN. The Shiré Highlands (East Central Africa). By JOHN BUCHANAN, Planter at Zomba. Crown 8vo, 5s.

BURBIDGE.
Domestic Floriculture, Window Gardening, and Floral Decorations. Being practical directions for the Propagation, Culture, and Arrangement of Plants and Flowers as Domestic Ornaments. By F. W. BURBIDGE. Second Edition. Crown 8vo, with numerous Illustrations, 7s. 6d.

Cultivated Plants: Their Propagation and Improvement. Including Natural and Artificial Hybridisation, Raising from Seed, Cuttings, and Layers, Grafting and Budding, as applied to the Families and Genera in Cultivation. Crown 8vo, with numerous Illustrations, 12s. 6d.

BURGESS. Ragnarök. A Tale of the White Christ. By J. J. HALDANE BURGESS, Author of 'Rasmie's Büidlie,' 'Shetland Sketches,' &c. Crown 8vo, 6s.

BURROWS. Commentaries on the History of England, from the Earliest Times to 1865. By MONTAGU BURROWS, Chichele Professor of Modern History in the University of Oxford; Captain R.N.; F.S.A., &c.; "Officier de l'Instruction Publique," France. Crown 8vo, 7s. 6d.

BURTON.
The History of Scotland: From Agricola's Invasion to the Extinction of the last Jacobite Insurrection. By JOHN HILL BURTON, D.C.L., Historiographer-Royal for Scotland. New and Enlarged Edition, 8 vols., and Index. Crown 8vo, £3, 3s.

History of the British Empire during the Reign of Queen Anne. In 3 vols. 8vo. 36s.

The Scot Abroad. Third Edition. Crown 8vo, 10s. 6d.

The Book-Hunter. New Edition. With Portrait. Crown 8vo, 7s. 6d.

BUTE.
The Roman Breviary: Reformed by Order of the Holy Œcumenical Council of Trent; Published by Order of Pope St Pius V.; and Revised by Clement VIII. and Urban VIII.; together with the Offices since granted. Translated out of Latin into English by JOHN, Marquess of Bute, K.T. In 2 vols. crown 8vo, cloth boards, edges uncut. £2, 2s.

The Altus of St Columba. With a Prose Paraphrase and Notes. In paper cover, 2s. 6d.

BUTT.
Miss Molly. By BEATRICE MAY BUTT. Cheap Edition, 2s.

Eugenie. Crown 8vo, 6s. 6d.

Elizabeth, and other Sketches. Crown 8vo, 6s.

Delicia. New Edition. Crown 8vo, 2s. 6d.

CAIRD.
Sermons. By JOHN CAIRD, D.D., Principal of the University of Glasgow. Seventeenth Thousand. Fcap. 8vo, 5s.

Religion in Common Life. A Sermon preached in Crathie Church, October 14, 1855, before Her Majesty the Queen and Prince Albert. Published by Her Majesty's Command. Cheap Edition, 3d.

CALDER. Chaucer's Canterbury Pilgrimage. Epitomised by WILLIAM CALDER. With Photogravure of the Pilgrimage Company, and other Illustrations, Glossary, &c. Crown 8vo, gilt edges, 4s. Cheaper Edition without Photogravure Plate. Crown 8vo, 2s. 6d.

CAMPBELL. Critical Studies in St Luke's Gospel: Its Demonology and Ebionitism. By COLIN CAMPBELL, D.D., Minister of the Parish of Dundee, formerly Scholar and Fellow of Glasgow University. Author of the 'Three First Gospels in Greek, arranged in parallel columns.' Post 8vo, 7s. 6d.

CAMPBELL. Sermons Preached before the Queen at Balmoral. By the Rev. A. A. CAMPBELL, Minister of Crathie. Published by Command of Her Majesty. Crown 8vo, 4s. 6d.

CAMPBELL. Records of Argyll. Legends, Traditions, and Recollections of Argyllshire Highlanders, collected chiefly from the Gaelic. With Notes on the Antiquity of the Dress, Clan Colours, or Tartans of the Highlanders. By Lord ARCHIBALD CAMPBELL. Illustrated with Nineteen full-page Etchings. 4to, printed on hand-made paper, £3, 3s.

CANTON. A Lost Epic, and other Poems. By WILLIAM CANTON. Crown 8vo, 5s.

CARRICK. Koumiss; or, Fermented Mare's Milk: and its uses in the Treatment and Cure of Pulmonary Consumption, and other Wasting Diseases. With an Appendix on the best Methods of Fermenting Cow's Milk. By GEORGE L. CARRICK, M.D., L.R.C.S.E. and L.R.C.P.E., Physician to the British Embassy, St Petersburg, &c. Crown 8vo, 10s. 6d.

CARSTAIRS. British Work in India. By R. CARSTAIRS. Crown 8vo, 6s.

CAUVIN. A Treasury of the English and German Languages. Compiled from the best Authors and Lexicographers in both Languages. By JOSEPH CAUVIN, LL.D. and Ph.D., of the University of Göttingen, &c. Crown 8vo, 7s. 6d.

CAVE-BROWNE. Lambeth Palace and its Associations. By J. CAVE-BROWNE, M.A., Vicar of Detling, Kent, and for many years Curate of Lambeth Parish Church. With an Introduction by the Archbishop of Canterbury. Second Edition, containing an additional Chapter on Medieval Life in the Old Palaces. 8vo, with Illustrations, 21s.

CHARTERIS. Canonicity; or, Early Testimonies to the Existence and Use of the Books of the New Testament. Based on Kirchhoffer's 'Quellensammlung.' Edited by A. H. CHARTERIS, D.D., Professor of Biblical Criticism in the University of Edinburgh. 8vo, 18s.

CHENNELLS. Recollections of an Egyptian Princess. By her English Governess (Miss E. CHENNELLS). Being a Record of Five Years' Residence at the Court of Ismael Pasha, Khédive. Second Edition. With Three Portraits. Post 8vo, 7s. 6d.

CHESNEY. The Dilemma. By General Sir GEORGE CHESNEY, K.C.B., M.P., Author of 'The Battle of Dorking,' &c. New Edition. Crown 8vo, 3s. 6d.

CHRISTISON. Life of Sir Robert Christison, Bart., M.D., D.C.L. Oxon., Professor of Medical Jurisprudence in the University of Edinburgh. Edited by his SONS. In 2 vols. 8vo. Vol. I.—Autobiography. 16s. Vol. II.—Memoirs. 16s.

CHRONICLES OF WESTERLY: A Provincial Sketch. By the Author of 'Culmshire Folk,' 'John Orlebar,' &c. 3 vols. crown 8vo, 25s. 6d.

CHURCH SERVICE SOCIETY.

A Book of Common Order: being Forms of Worship issued by the Church Service Society. Sixth Edition. Crown 8vo, 6s. Also in 2 vols. crown 8vo, 6s. 6d.

Daily Offices for Morning and Evening Prayer throughout the Week. Crown 8vo, 3s. 6d.

Order of Divine Service for Children. Issued by the Church Service Society. With Scottish Hymnal. Cloth, 3d.

CLOUSTON. Popular Tales and Fictions: their Migrations and Transformations. By W. A. CLOUSTON, Editor of 'Arabian Poetry for English Readers,' &c. 2 vols. post 8vo, roxburghe binding, 25s.

COCHRAN. A Handy Text-Book of Military Law. Compiled chiefly to assist Officers preparing for Examination; also for all Officers of the Regular and Auxiliary Forces. Comprising also a Synopsis of part of the Army Act. By Major F. COCHRAN, Hampshire Regiment Garrison Instructor, North British District. Crown 8vo, 7s. 6d.

COLQUHOUN. The Moor and the Loch. Containing Minute Instructions in all Highland Sports, with Wanderings over Crag and Corrie, Flood and Fell. By JOHN COLQUHOUN. Cheap Edition. With Illustrations. Demy 8vo, 10s. 6d.

COLVILE. Round the Black Man's Garden. By ZÉLIE COLVILE, F.R.G.S. With 2 Maps and 50 Illustrations from Drawings by the Author and from Photographs. Demy 8vo, 16s.

CONSTITUTION AND LAW OF THE CHURCH OF SCOTLAND. With an Introductory Note by the late Principal Tulloch. New Edition, Revised and Enlarged. Crown 8vo, 3s. 6d.

COTTERILL. Suggested Reforms in Public Schools. By C. C.
COTTERILL, M.A. Crown 8vo, 3s. 6d.

CRANSTOUN.
The Elegies of Albius Tibullus. Translated into English
Verse, with Life of the Poet, and Illustrative Notes. By JAMES CRANSTOUN,
LL.D., Author of a Translation of 'Catullus.' Crown 8vo, 6s. 6d.

The Elegies of Sextus Propertius. Translated into English
Verse, with Life of the Poet, and Illustrative Notes. Crown 8vo, 7s. 6d.

CRAWFORD. An Atonement of East London, and other Poems.
By HOWARD CRAWFORD, M.A. Crown 8vo, 5s.

CRAWFORD. Saracinesca. By F. MARION CRAWFORD, Author
of 'Mr Isaacs,' &c. &c. Eighth Edition. Crown 8vo, 6s.

CRAWFORD.
The Doctrine of Holy Scripture respecting the Atonement.
By the late THOMAS J. CRAWFORD, D.D., Professor of Divinity in the University
of Edinburgh. Fifth Edition. 8vo, 12s.

The Fatherhood of God, Considered in its General and Special
Aspects. Third Edition, Revised and Enlarged. 8vo, 9s.

The Preaching of the Cross, and other Sermons. 8vo, 7s. 6d.

The Mysteries of Christianity. Crown 8vo, 7s. 6d.

CROSS. Impressions of Dante, and of the New World ; with a
Few Words on Bimetallism. By J. W. CROSS, Editor of 'George Eliot's Life, as
related in her Letters and Journals.' Post 8vo, 6s.

CUSHING.
The Blacksmith of Voe. By PAUL CUSHING, Author of 'The
Bull i' th' Thorn,' 'Cut with his own Diamond.' Cheap Edition. Crown 8vo, 3s. 6d.

DAVIES.
Norfolk Broads and Rivers ; or, The Waterways, Lagoons,
and Decoys of East Anglia. By G. CHRISTOPHER DAVIES. Illustrated with
Seven full-page Plates. New and Cheaper Edition. Crown 8vo, 6s.

Our Home in Aveyron. Sketches of Peasant Life in Aveyron
and the Lot. By G. CHRISTOPHER DAVIES and Mrs BROUGHALL. Illustrated
with full-page Illustrations. 8vo, 15s. Cheap Edition, 7s. 6d.

DE LA WARR. An Eastern Cruise in the 'Edeline.' By the
Countess DE LA WARR. In Illustrated Cover. 2s.

DESCARTES. The Method, Meditations, and Principles of Philo-
sophy of Descartes. Translated from the Original French and Latin. With a
New Introductory Essay, Historical and Critical, on the Cartesian Philosophy.
By Professor VEITCH, LL.D., Glasgow University. Tenth Edition. 6s. 6d.

DEWAR. Voyage of the "Nyanza," R.N.Y.C. Being the Record
of a Three Years' Cruise in a Schooner Yacht in the Atlantic and Pacific, and her
subsequent Shipwreck. By J. CUMMING DEWAR, late Captain King's Dragoon
Guards and 11th Prince Albert's Hussars. With Two Autogravures, numerous
Illustrations, and a Map. Demy 8vo, 21s.

DICKSON. Gleanings from Japan. By W. G. DICKSON, Author
of 'Japan : Being a Sketch of its History, Government, and Officers of the
Empire.' With Illustrations. 8vo, 16s.

DOGS, OUR DOMESTICATED : Their Treatment in reference
to Food, Diseases, Habits, Punishment, Accomplishments. By 'MAGENTA.'
Crown 8vo, 2s. 6d.

DOUGLAS. Chinese Stories. By ROBERT K. DOUGLAS. With numerous Illustrations by Parkinson, Forestier, and others. New and Cheaper Edition. Small demy 8vo, 5s.

DU CANE. The Odyssey of Homer, Books I.-XII. Translated into English Verse. By Sir CHARLES DU CANE, K.C.M.G. 8vo, 10s. 6d.

DUDGEON. History of the Edinburgh or Queen's Regiment Light Infantry Militia, now 3rd Battalion The Royal Scots; with an Account of the Origin and Progress of the Militia, and a Brief Sketch of the Old Royal Scots. By Major R. C. DUDGEON, Adjutant 3rd Battalion the Royal Scots. Post 8vo, with Illustrations, 10s. 6d.

DUNCAN. Manual of the General Acts of Parliament relating to the Salmon Fisheries of Scotland from 1828 to 1882. By J. BARKER DUNCAN. Crown 8vo, 5s.

DUNN. Red Cap and Blue Jacket: A Novel. By GEORGE DUNN. 3 vols. crown 8vo, 25s. 6d.

DUNSMORE. Manual of the Law of Scotland as to the Relations between Agricultural Tenants and the Landlords, Servants, Merchants, and Bowers. By W. DUNSMORE. 8vo, 7s. 6d.

DUPRÈ. Thoughts on Art, and Autobiographical Memoirs of Giovanni Dupre. Translated from the Italian by E. M. PERUZZI, with the permission of the Author. New Edition. With an Introduction by W. W. STORY. Crown 8vo, 10s. 6d.

ELIOT.
George Eliot's Life, Related in Her Letters and Journals. Arranged and Edited by her husband, J. W. Cross. With Portrait and other Illustrations. Third Edition. 3 vols. post 8vo, 42s.

George Eliot's Life. (Cabinet Edition.) With Portrait and other Illustrations. 3 vols. crown 8vo, 15s.

George Eliot's Life. With Portrait and other Illustrations. New Edition, in one volume. Crown 8vo, 7s. 6d.

Works of George Eliot (Cabinet Edition). 21 volumes, crown 8vo, price £5, 5s. Also to be had handsomely bound in half and full calf. The Volumes are sold separately, bound in cloth, price 5s. each—viz.: Romola. 2 vols.—Silas Marner, The Lifted Veil, Brother Jacob. 1 vol.—Adam Bede. 2 vols.—Scenes of Clerical Life. 2 vols.—The Mill on the Floss. 2 vols.—Felix Holt. 2 vols.—Middlemarch. 3 vols.—Daniel Deronda. 3 vols.—The Spanish Gypsy. 1 vol.—Jubal, and other Poems, Old and New. 1 vol.—Theophrastus Such. 1 vol.—Essays. 1 vol.

Novels by George Eliot. Cheap Edition. Adam Bede. Illustrated. 3s. 6d., cloth.—The Mill on the Floss. Illustrated. 3s. 6d., cloth.—Scenes of Clerical Life. Illustrated. 3s., cloth.—Silas Marner: the Weaver of Raveloe. Illustrated. 2s. 6d., cloth.—Felix Holt, the Radical. Illustrated. 3s. 6d., cloth.—Romola. With Vignette. 3s. 6d., cloth.

Middlemarch. Crown 8vo, 7s. 6d.

Daniel Deronda. Crown 8vo, 7s. 6d.

Essays. New Edition. Crown 8vo, 5s.

Impressions of Theophrastus Such. New Edition. Crown 8vo, 5s.

The Spanish Gypsy. New Edition. Crown 8vo, 5s.

The Legend of Jubal, and other Poems, Old and New. New Edition. Crown 8vo, 5s.

Wise, Witty, and Tender Sayings, in Prose and Verse. Selected from the Works of GEORGE ELIOT. New Edition. Fcap. 8vo, 3s. 6d.

ELIOT.

The George Eliot Birthday Book. Printed on fine paper, with red border, and handsomely bound in cloth, gilt. Fcap. 8vo, 3s. 6d. And In French morocco or Russia, 5s.

ESSAYS ON SOCIAL SUBJECTS. Originally published in the 'Saturday Review.' New Edition. First and Second Series. 2 vols. crown 8vo, 6s. each.

FAITHS OF THE WORLD, The. A Concise History of the Great Religious Systems of the World. By various Authors. Crown 8vo, 5s!

FARRER. A Tour in Greece in 1880. By RICHARD RIDLEY FARRER. With Twenty-seven full-page Illustrations by Lord WINDSOR. Royal 8vo, with a Map, 21s.

FERRIER.

Philosophical Works of the late James F. Ferrier, B.A. Oxon., Professor of Moral Philosophy and Political Economy, St Andrews. New Edition. Edited by Sir ALEXANDER GRANT, Bart., D.C.L., and Professor LUSHINGTON. 3 vols. crown 8vo, 34s. 6d.

Institutes of Metaphysic. Third Edition. 10s. 6d.

Lectures on the Early Greek Philosophy. 4th Edition. 10s. 6d.

Philosophical Remains, including the Lectures on Early Greek Philosophy. New Edition. 2 vols. 24s.

FITZROY. Dogma and the Church of England. By A. I. FITZROY. Post 8vo, 7s. 6d.

FLINT.

Historical Philosophy in France and French Belgium and Switzerland. By ROBERT FLINT, Corresponding Member of the Institute of France, Hon. Member of the Royal Society of Palermo, Professor in the University of Edinburgh, &c. 8vo, 21s.

Agnosticism. Being the Croall Lecture for 1887-88.
[*In the press.*

Theism. Being the Baird Lecture for 1876. Eighth Edition, Revised. Crown 8vo, 7s. 6d.

Anti-Theistic Theories. Being the Baird Lecture for 1877. Fifth Edition. Crown 8vo, 10s. 6d.

FOREIGN CLASSICS FOR ENGLISH READERS. Edited by Mrs OLIPHANT. Price 2s. 6d *For List of Volumes, see page 2.*

FOSTER. The Fallen City, and other Poems. By WILL FOSTER. Crown 8vo, 6s.

FRANCILLON. Gods and Heroes; or, The Kingdom of Jupiter. By R. E. FRANCILLON. With 8 Illustrations. Crown 8vo, 5s.

FROM SPRING TO FALL; OR, WHEN LIFE STIRS. By "A SON OF THE MARSHES.' *See page 28.*

FULLARTON.

Merlin: A Dramatic Poem. By RALPH MACLEOD FULLARTON. Crown 8vo, 5s.

Tanhäuser. Crown 8vo, 6s.

Lallan Sangs and German Lyrics. Crown 8vo, 5s.

GALT. Novels by JOHN GALT. Fcap. 8vo, boards, each 2s.; cloth, 2s. 6d.
ANNALS OF THE PARISH.—THE PROVOST.—SIR ANDREW WYLIE.—THE ENTAIL.

GENERAL ASSEMBLY OF THE CHURCH OF SCOTLAND.

Scottish Hymnal, With Appendix Incorporated. Published for use in Churches by Authority of the General Assembly. 1. Large type, cloth, red edges, 2s. 6d.; French morocco, 4s. 2. Bourgeois type, limp cloth, 1s.; French morocco, 2s. 3. Nonpareil type, cloth, red edges, 6d.; French morocco, 1s. 4d. 4. Paper covers, 3d. 5. Sunday-School Edition, paper covers, 1d., cloth, 2d. No. 1, bound with the Psalms and Paraphrases, French morocco, 8s. No. 2, bound with the Psalms and Paraphrases, cloth, 2s.; French morocco, 3s.

Prayers for Social and Family Worship. Prepared by a Special Committee of the General Assembly of the Church of Scotland. Entirely New Edition, Revised and Enlarged. Fcap. 8vo, red edges, 2s.

Prayers for Family Worship. A Selection of Four Weeks' Prayers. New Edition. Authorised by the General Assembly of the Church of Scotland Fcap. 8vo, red edges, 1s. 6d.

One Hundred Prayers. Prepared by a Committee of the General Assembly of the Church of Scotland. 16mo, cloth limp. [*In preparation.*

GERARD.

Reata : What's in a Name. By E. D. GERARD. Cheap Edition. Crown 8vo, 3s. 6d.

Beggar my Neighbour. Cheap Edition. Crown 8vo, 3s. 6d.

The Waters of Hercules. Cheap Edition. Crown 8vo, 3s. 6d.

A Sensitive Plant. Crown 8vo, 3s. 6d.

GERARD.

The Land beyond the Forest. Facts, Figures, and Fancies from Transylvania. By E. GERARD. With Maps and Illustrations. 2 vols. post 8vo, 25s.

Bis : Some Tales Retold. Crown 8vo, 6s.

A Secret Mission. 2 vols. crown 8vo, 17s.

GERARD.

Lady Baby. By DOROTHEA GERARD. Cheap Edition. Crown 8vo, 3s. 6d.

Recha. Second Edition. Crown 8vo, 6s.

The Rich Miss Riddell. Crown 8vo, 6s.

GERARD. Stonyhurst Latin Grammar. By Rev. JOHN GERARD. Second Edition. Fcap. 8vo, 3s.

GILL.

Free Trade : an Inquiry into the Nature of its Operation. By RICHARD GILL. Crown 8vo, 7s. 6d.

Free Trade under Protection. Crown 8vo, 7s. 6d.

GOETHE. Poems and Ballads of Goethe. Translated by Professor AYTOUN and Sir THEODORE MARTIN, K.C.B. Third Edition. Fcap. 8vo, 6s.

GOETHE'S FAUST. Translated into English Verse by Sir THEODORE MARTIN, K.C.B. Part I. Second Edition, crown 8vo, 6s. Ninth Edition, fcap., 3s. 6d. Part II. Second Edition, Revised. Fcap. 8vo, 6s.

GORDON CUMMING.

At Home in Fiji. By C. F. GORDON CUMMING. Fourth Edition, post 8vo. With Illustrations and Map. 7s. 6d.

A Lady's Cruise in a French Man-of-War. New and Cheaper Edition. 8vo. With Illustrations and Map. 12s. 6d.

Fire-Fountains. The Kingdom of Hawaii : Its Volcanoes, and the History of its Missions. With Map and Illustrations. 2 vols. 8vo, 25s.

Wanderings in China. New and Cheaper Edition. 8vo, with Illustrations, 10s.

Granite Crags : The Yo-semité Region of California. Illustrated with 8 Engravings. New and Cheaper Edition. 8vo, 8s. 6d.

GRAHAM. The Life and Work of Syed Ahmed Khan, C.S.I. By Lieut.-Colonel G. F. I. GRAHAM, B.S.C. 8vo, 14s.

GRAHAM. **Manual of the Elections (Scot.) (Corrupt and Illegal**
Practices) Act, 1890. With Analysis, Relative Act of Sederunt, Appendix containing the Corrupt Practices Acts of 1883 and 1885, and Copious Index. By J.
EDWARD GRAHAM, Advocate. 8vo, 4s. 6d.

GRAND.
A Domestic Experiment. By SARAH GRAND, Author of
'The Heavenly Twins,' 'Ideala: A Study from Life.' Crown 8vo, 6s.
Singularly Deluded. Crown 8vo, 6s.

GRANT. **Bush-Life in Queensland.** By A. C. GRANT. New
Edition. Crown 8vo, 6s.

GRANT. **Life of Sir Hope Grant.** With Selections from his
Correspondence. Edited by HENRY KNOLLYS, Colonel (H.P.) Royal Artillery,
his former A.D.C., Editor of 'Incidents in the Sepoy War;' Author of 'Sketches
of Life in Japan,' &c. With Portraits of Sir Hope Grant and other Illustrations. Maps and Plans. 2 vols. demy 8vo, 21s.

GRIER. **In Furthest Ind.** The Narrative of Mr EDWARD CAR-
LYON of Ellswether, in the County of Northampton, and late of the Honourable
East India Company's Service, Gentleman. Wrote by his own hand in the year
of grace 1697. Edited, with a few Explanatory Notes, by SYDNEY C. GRIER.
Post 8vo, 6s.

GUTHRIE-SMITH. **Crispus: A Drama.** By H. GUTHRIE-
SMITH. Fcap. 4to, 5s.

HALDANE. **Subtropical Cultivations and Climates.** A Handy
Book for Planters, Colonists, and Settlers. By R. C. HALDANE. Post 8vo, 9s.

HAMERTON.
Wenderholme: A Story of Lancashire and Yorkshire Life.
By P. G. HAMERTON, Author of 'A Painter's Camp.' Crown 8vo, 6s.
Marmorne. New Edition. Crown 8vo, 3s. 6d.

HAMILTON.
Lectures on Metaphysics. By Sir WILLIAM HAMILTON,
Bart., Professor of Logic and Metaphysics in the University of Edinburgh.
Edited by the Rev. H. L. MANSEL, B.D., LL.D., Dean of St Paul's; and JOHN
VEITCH, M.A., LL.D., Professor of Logic and Rhetoric, Glasgow. Seventh
Edition. 2 vols. 8vo, 24s.
Lectures on Logic. Edited by the SAME. Third Edition,
Revised. 2 vols., 24s.
Discussions on Philosophy and Literature, Education and
University Reform. Third Edition. 8vo, 21s.
Memoir of Sir William Hamilton, Bart., Professor of Logic
and Metaphysics in the University of Edinburgh. By Professor VEITCH, of the
University of Glasgow. 8vo, with Portrait, 18s.
Sir William Hamilton: The Man and his Philosophy. Two
Lectures delivered before the Edinburgh Philosophical Institution, January and
February 1883. By Professor VEITCH. Crown 8vo, 2s.

HAMLEY.
The Operations of War Explained and Illustrated. By
General Sir EDWARD BRUCE HAMLEY, K.C.B., K.C.M.G. Fifth Edition, Revised
throughout. 4to, with numerous Illustrations, 30s.
National Defence; Articles and Speeches. Post 8vo, 6s.
Shakespeare's Funeral, and other Papers. Post 8vo, 7s 6d.
Thomas Carlyle: An Essay. Second Edition. Crown 8vo,
2s. 6d.
On Outposts. Second Edition. 8vo, 2s.
Wellington's Career; A Military and Political Summary.
Crown 8vo, 2s.

HAMLEY.
Lady Lee's Widowhood. New Edition. Crown 8vo, 3s. 6d.
Cheaper Edition, 2s. 6d.
Our Poor Relations. A Philozoic Essay. With Illustrations, chiefly by Ernest Griset. Crown 8vo, cloth gilt, 3s. 6d.

HARRADEN.
In Varying Moods : Short Stories. By BEATRICE HARRADEN, Author of 'Ships that Pass in the Night.' Ninth Edition. Crown 8vo, 3s. 6d.

HARRIS.
A Journey through the Yemen, and some General Remarks upon that Country. By WALTER B. HARRIS, F.R.G.S., Author of 'The Land of an African Sultan ; Travels in Morocco,' &c. With 3 Maps and numerous Illustrations by Forestier and Wallace from Sketches and Photographs taken by the Author. Demy 8vo, 16s.

HAWKER.
The Prose Works of Rev. R. S. HAWKER, Vicar of Morwenstow. Including 'Footprints of Former Men in Far Cornwall.' Re-edited, with Sketches never before published. With a Frontispiece. Crown 8vo, 3s. 6d.

HAY.
The Works of the Right Rev. Dr George Hay, Bishop of Edinburgh. Edited under the Supervision of the Right Rev. Bishop STRAIN. With Memoir and Portrait of the Author. 5 vols. crown 8vo, bound in extra cloth, £1, 1s. The following Volumes may be had separately—viz. :
The Devout Christian Instructed in the Law of Christ from the Written Word. 2 vols., 8s.—The Pious Christian Instructed in the Nature and Practice of the Principal Exercises of Piety. 1 vol., 3s.

HEATLEY.
The Horse-Owner's Safeguard. A Handy Medical Guide for every Man who owns a Horse. By G. S. HEATLEY, M.R.C.V.S. Crown 8vo, 5s.
The Stock-Owner's Guide. A Handy Medical Treatise for every Man who owns an Ox or a Cow. Crown 8vo, 4s. 6d.

HEDDERWICK.
Lays of Middle Age ; and other Poems. By JAMES HEDDERWICK, LL.D., Author of 'Backward Glances.' Price 3s. 6d.

HEMANS.
The Poetical Works of Mrs Hemans. Copyright Editions. Royal 8vo, 5s. The Same with Engravings, cloth, gilt edges, 7s. 6d.
Select Poems of Mrs Hemans. Fcap., cloth, gilt edges, 3s.

HERKLESS.
Cardinal Beaton : Priest and Politician. By JOHN HERKLESS, Professor of Church History, St Andrews. With a Portrait. Post 8vo, 7s. 6d.

HEWISON.
The Isle of Bute in the Olden Time. With Illustrations, Maps, and Plans. By JAMES KING HEWISON, M.A., F.S.A. (Scot.), Minister of Rothesay. Vol. I., Celtic Saints and Heroes. Crown 4to, 15s. net.
[*Vol II. in preparation.*

HOME PRAYERS.
By Ministers of the Church of Scotland and Members of the Church Service Society. Second Edition. Fcap. 8vo, 3s.

HOMER.
The Odyssey. Translated into English Verse in the Spenserian Stanza. By PHILIP STANHOPE WORSLEY. 3d Edition. 2 vols. fcap., 12s.
The Iliad. Translated by P. S. WORSLEY and Professor CONINGTON. 2 vols. crown 8vo, 21s.

HUTCHINSON.
Hints on the Game of Golf. By HORACE G. HUTCHINSON. Eighth Edition, Enlarged. Fcap. 8vo, cloth, 1s.

IDDESLEIGH.
Lectures and Essays. By the late EARL of IDDESLEIGH, G.C.B., D.C.L., &c. 8vo, 16s.
Life, Letters, and Diaries of Sir Stafford Northcote, First Earl of Iddesleigh. By ANDREW LANG. With Three Portraits and a View of Pynes. Third Edition. 2 vols. post 8vo, 31s. 6d.
POPULAR EDITION. With Portrait and View of Pynes. Post 8vo, 7s. 6d.

INDEX GEOGRAPHICUS: Being a List, alphabetically arranged, of the Principal Places on the Globe, with the Countries and Subdivisions of the Countries in which they are situated, and their Latitudes and Longitudes. Imperial 8vo, pp. 676, 21s.

JEAN JAMBON. Our Trip to Blunderland ; or, Grand Excursion to Blundertown and Back. By JEAN JAMBON. With Sixty Illustrations designed by CHARLES DOYLE, engraved by DALZIEL. Fourth Thousand. Cloth, gilt edges, 6s. 6d. Cheap Edition, cloth, 3s. 6d. Boards, 2s. 6d.

JEBB. A Strange Career. The Life and Adventures of JOHN GLADWYN JEBB. By his Widow. With an Introduction by H. RIDER HAGGARD, and an Electrogravure Portrait of Mr Jebb. Small demy 8vo, 10s. 6d.

JENNINGS. Mr Gladstone : A Study. By LOUIS J. JENNINGS, M.P., Author of 'Republican Government in the United States,' 'The Croker Memoirs,' &c. Popular Edition. Crown 8vo, 1s.

JERNINGHAM.
Reminiscences of an Attaché. By HUBERT E. H. JERNINGHAM. Second Edition. Crown 8vo, 5s.
Diane de Breteuille. A Love Story. Crown 8vo, 2s. 6d.

JOHNSTON.
The Chemistry of Common Life. By Professor J. F. W. JOHNSTON. New Edition, Revised. By ARTHUR HERBERT CHURCH, M.A. Oxon.; Author of 'Food : its Sources, Constituents, and Uses,' &c. With Maps and 102 Engravings. Crown 8vo, 7s. 6d.
Elements of Agricultural Chemistry. An entirely New Edition from the Edition by Sir CHARLES A. CAMERON, M.D., F.R.C.S.I., &c. Revised and brought down to date by C. M. AIKMAN, M.A., B.Sc., F.R.S.E., Professor of Chemistry, Glasgow Veterinary College. Crown 8vo, 6s. 6d.
Catechism of Agricultural Chemistry. An entirely New Edition from the Edition by Sir CHARLES A. CAMERON. Revised and Enlarged by C. M. AIKMAN, M.A., &c. 92d Thousand. With numerous Illustrations. Crown 8vo, 1s.

JOHNSTON. Agricultural Holdings (Scotland) Acts, 1883 and 1889 ; and the Ground Game Act, 1880. With Notes, and Summary of Procedure, &c. By CHRISTOPHER N. JOHNSTON, M.A., Advocate. Demy 8vo, 5s.

JOKAI. Timar's Two Worlds. By MAURUS JOKAI. Authorised Translation by Mrs HEGAN KENNARD. Cheap Edition. Crown 8vo, 6s.

KEBBEL. The Old and the New : English Country Life. By T. E. KEBBEL, M.A., Author of 'The Agricultural Labourers,'·'Essays in History and Politics,' 'Life of Lord Beaconsfield.' Crown 8vo, 5s.

KING. The Metamorphoses of Ovid. Translated in English Blank Verse. By HENRY KING, M.A., Fellow of Wadham College, Oxford, and of the Inner Temple, Barrister-at-Law. Crown 8vo, 10s. 6d.

KINGLAKE.
History of the Invasion of the Crimea. By A. W. KINGLAKE. Cabinet Edition, Revised. With an Index to the Complete Work. Illustrated with Maps and Plans. Complete in 9 vols., crown 8vo, at 6s. each.
History of the Invasion of the Crimea. Demy 8vo. Vol. VI. Winter Troubles. With a Map, 16s. Vols. VII. and VIII. From the Morrow of Inkerman to the Death of Lord Raglan. With an Index to the Whole Work. With Maps and Plans. 28s.
Eothen. A New Edition, uniform with the Cabinet Edition of the 'History of the Invasion of the Crimea.' 6s.

KLEIN. Among the Gods. Scenes of India, with Legends by the Way. By AUGUSTA KLEIN. With Illustrations. In 1 vol. demy 8vo.
[*In the press.*]

KNEIPP. My Water-Cure. As Tested through more than
Thirty Years, and Described for the Healing of Diseases and the Preservation of
Health. By SEBASTIAN KNEIPP, Parish Priest of Wörishofen (Bavaria). With a
Portrait and other Illustrations. Authorised English Translation from the
Thirtieth German Edition, by A. de F. Cheap Edition. With an Appendix, con-
taining the Latest Developments of Pfarrer Kneipp's System, and a Preface by
E. Gerard. Crown 8vo, 3s. 6d.

KNOLLYS. The Elements of Field-Artillery. Designed for
the Use of Infantry and Cavalry Officers. By HENRY KNOLLYS, Colonel Royal
Artillery; Author of 'From Sedan to Saarbrück,' Editor of 'Incidents in the
Sepoy War,' &c. With Engravings. Crown 8vo, 7s. 6d.

LAMINGTON. In the Days of the Dandies. By the late Lord
LAMINGTON. Crown 8vo. Illustrated cover, 1s.; cloth, 1s. 6d.

LANG. Life, Letters, and Diaries of Sir Stafford Northcote,
First Earl of Iddesleigh. By ANDREW LANG. With Three Portraits and a View
of Pynes. Third Edition. 2 vols. post 8vo, 31s. 6d.
POPULAR EDITION. With Portrait and View of Pynes. Post 8vo, 7s. 6d.

LAWLESS. Hurrish : A Study. By the Hon. EMILY LAWLESS,
Author of 'A Chelsea Householder,' &c. Fourth Edition. Crown 8vo, 3s. 6d.

LEES. A Handbook of the Sheriff and Justice of Peace Small
Debt Courts. With Notes, References, and Forms. By J. M. LEES, Advocate,
Sheriff of Stirling, Dumbarton, and Clackmannan. 8vo, 7s. 6d.

LINDSAY. The Progressiveness of Modern Christian Thought.
By the Rev. JAMES LINDSAY, M.A., B.D., B.Sc., F.R.S.E., F.G.S., Minister of
the Parish of St Andrew's, Kilmarnock. Crown 8vo, 6s.

LLOYD. Ireland under the Land League. A Narrative of
Personal Experiences. By CLIFFORD LLOYD, Special Resident Magistrate.
Post 8vo, 6s.

LOCKHART.
Doubles and Quits. By LAURENCE W. M. LOCKHART. New
Edition. Crown 8vo, 3s. 6d.
Fair to See. New Edition. Crown 8vo, 3s. 6d.
Mine is Thine. New Edition. Crown 8vo, 3s. 6d.

LOCKHART.
The Church of Scotland in the Thirteenth Century. The
Life and Times of David de Bernham of St Andrews (Bishop), A.D. 1239 to 1253.
With List of Churches dedicated by him, and Dates. By WILLIAM LOCKHART,
A.M., D.D., F.S.A. Scot., Minister of Colinton Parish. 2d Edition. 8vo, 6s.
Dies Tristes : Sermons for Seasons of Sorrow. Crown 8vo, 6s.

LORIMER.
The Institutes of Law : A Treatise of the Principles of Juris-
prudence as determined by Nature. By the late JAMES LORIMER, Professor of
Public Law and of the Law of Nature and Nations in the University of Edin-
burgh. New Edition, Revised and much Enlarged. 8vo, 18s.
The Institutes of the Law of Nations. A Treatise of the
Jural Relation of Separate Political Communities. In 2 vols. 8vo. Volume I.,
price 16s. Volume II., price 20s.

LOVE. Scottish Church Music. Its Composers and Sources.
With Musical Illustrations. By JAMES LOVE. Post 8vo, 7s. 6d.

LUGARD. The Rise of our East African Empire : Early Efforts
in Uganda and Nyasaland. By F. D. LUGARD, Captain Norfolk Regiment.
With 130 Illustrations from Drawings and Photographs under the personal
superintendence of the Author, and 14 specially prepared Maps. In 2 vols. large
demy 8vo, 42s.

M'COMBIE. Cattle and Cattle-Breeders. By WILLIAM M'COMBIE,
Tillyfour. New Edition, Enlarged, with Memoir of the Author by JAMES
MACDONALD, F.R.S.E., Secretary Highland and Agricultural Society of Scotland.
Crown 8vo, 3s. 6d.

M'CRIE.
Works of the Rev. Thomas M'Crie, D.D. Uniform Edition.
4 vols. crown 8vo, 24s.
Life of John Knox. Crown 8vo, 6s. Another Edition, 3s. 6d.
Life of Andrew Melville. Crown 8vo, 6s.
History of the Progress and Suppression of the Reformation
in Italy in the Sixteenth Century. Crown 8vo, 4s.
History of the Progress and Suppression of the Reformation
in Spain in the Sixteenth Century. Crown 8vo, 3s. 6d.
Lectures on the Book of Esther. Fcap. 8vo, 5s.

M'CRIE. The Public Worship of Presbyterian Scotland. Histori-
cally treated. With copious Notes, Appendices, and Index. The Fourteenth
Series of the Cunningham Lectures. By the Rev. CHARLES G. M'CRIE, D.D.
Demy 8vo, 10s. 6d.

MACDONALD. A Manual of the Criminal Law (Scotland) Procedure Act, 1887. By NORMAN DORAN MACDONALD. Revised by the LORD
JUSTICE-CLERK. 8vo, 10s. 6d.

MACDONALD.
Stephens' Book of the Farm. Fourth Edition. Revised and
in great part Rewritten by JAMES MACDONALD, F.R.S.E., Secretary, Highland
and Agricultural Society of Scotland. Complete in 3 vols., bound with leather
back, gilt top, £3, 3s. In Six Divisional Vols., bound in cloth, each 10s. 6d.
Pringle's Live Stock of the Farm. Third Edition. Revised
and Edited by JAMES MACDONALD. Crown 8vo, 7s. 6d.
M'Combie's Cattle and Cattle - Breeders. New Edition,
Enlarged, with Memoir of the Author by JAMES MACDONALD. Crown 8vo, 3s. 6d.
History of Polled Aberdeen and Angus Cattle. Giving an
Account of the Origin, Improvement, and Characteristics of the Breed. By JAMES
MACDONALD and JAMES SINCLAIR. Illustrated with numerous Animal Portraits.
Post 8vo, 12s. 6d.

MACDOUGALL AND DODDS. A Manual of the Local Govern-
ment (Scotland) Act, 1894. With Introduction, Explanatory Notes, and Copious
Index. By J. PATTEN MACDOUGALL, Legal Secretary to the Lord Advocate, and
J. M. DODDS. Crown 8vo, 2s. 6d. net.

M'INTOSH. The Book of the Garden. By CHARLES M'INTOSH,
formerly Curator of the Royal Gardens of his Majesty the King of the Belgians,
and lately of those of his Grace the Duke of Buccleuch, K.G., at Dalkeith Palace.
2 vols. royal 8vo, with 1350 Engravings. £4, 7s. 6d. Vol. I. On the Formation
of Gardens and Construction of Garden Edifices, £2, 10s. Vol. II. Practical
Gardening, £1, 17s. 6d.

MACINTYRE. Hindu - Koh : Wanderings and Wild Sports on
and beyond the Himalayas. By Major-General DONALD MACINTYRE, V.C., late
Prince of Wales' Own Goorkhas, F.R.G.S. *Dedicated to H.R.H. The Prince of
Wales.* New and Cheaper Edition, Revised, with numerous Illustrations. Post
8vo, 3s. 6d.

MACKAY. A Sketch of the History of Fife and Kinross. A
Study of Scottish History and Character. By Æ. J. G. MACKAY, Sheriff of these
Counties. Crown 8vo, 6s.

MACKAY.
A Manual of Modern Geography ; Mathematical, Physical,
and Political. By the Rev. ALEXANDER MACKAY, LL.D., F.R.G.S. 11th
Thousand, Revised to the present time. Crown 8vo, pp. 688, 7s. 6d.
Elements of Modern Geography. 55th Thousand, Revised to
the present time. Crown 8vo, pp. 300, 3s.
The Intermediate Geography. Intended as an Intermediate
Book between the Author's 'Outlines of Geography' and 'Elements of Geo-
graphy.' Seventeenth Edition, Revised. Crown 8vo, pp. 238, 2s.

MACKAY.

Outlines of Modern Geography. 191st Thousand, Revised to
the present time. 18mo, pp. 128, 1s.

First Steps in Geography. 105th Thousand. 18mo, pp. 56.
Sewed, 4d.; cloth, 6d.

Elements of Physiography and Physical Geography. With
Express Reference to the Instructions issued by the Science and Art Department. 30th Thousand, Revised. Crown 8vo, 1s. 6d.

Facts and Dates : or, The Leading Events in Sacred and Profane History. and the Principal Facts in the various Physical Sciences. For
Schools and Private Reference. New Edition. Crown 8vo, 3s. 6d.

MACKENZIE.

Studies in Roman Law. With Comparative
Views of the Laws of France, England, and Scotland. By Lord MACKENZIE,
one of the Judges of the Court of Session in Scotland. Sixth Edition, Edited
by JOHN KIRKPATRICK, M.A., LL.B., Advocate, Professor of History in the
University of Edinburgh. 8vo, 12s.

MACPHERSON.

**Glimpses of Church and Social Life in the
Highlands in Olden Times.** By ALEXANDER MACPHERSON, F.S.A. Scot. With
6 Photogravure Portraits and other full-page Illustrations. Small 4to, 25s.

M'PHERSON.

Summer Sundays in a Strathmore Parish. By J. GORDON
M'PHERSON, Ph.D., F.R.S.E., Minister of Ruthven. Crown 8vo, 5s.

Golf and Golfers. Past and Present. With an Introduction
by the Right Hon. A. J. BALFOUR, and a Portrait of the Author. Fcap. 8vo,
1s. 6d.

MACRAE.

A Handbook of Deer-Stalking. By ALEXANDER
MACRAE, late Forester to Lord Henry Bentinck. With Introduction by Horatio
Ross, Esq. Fcap. 8vo, with 2 Photographs from Life. 3s. 6d.

MAIN.

Three Hundred English Sonnets. Chosen and Edited
by DAVID M. MAIN. Fcap. 8vo, 6s.

MAIR.

**A Digest of Laws and Decisions, Ecclesiastical and
Civil,** relating to the Constitution, Practice, and Affairs of the Church of Scotland. With Notes and Forms of Procedure. By the Rev. WILLIAM MAIR, D.D.,
Minister of the Parish of Earlston. Crown 8vo. [*New Edition in preparation.*]

MARCHMONT AND THE HUMES OF POLWARTH. By
One of their Descendants. With numerous Portraits and other Illustrations.
Crown 4to, 21s. net.

MARSHALL.

It Happened Yesterday. A Novel. By FREDERICK
MARSHALL, Author of 'Claire Brandon,' 'French Home Life.' Crown 8vo, 6s.

MARSHMAN.

History of India. From the Earliest Period to
the present time. By JOHN CLARK MARSHMAN, C.S.I. Third and Cheaper
Edition. Post 8vo, with Map, 6s.

MARTIN.

Goethe's Faust. Part I. Translated by Sir THEODORE MARTIN,
K.C.B. Second Edition, crown 8vo, 6s. Ninth Edition, fcap. 8vo, 3s. 6d.

Goethe's Faust. Part II. Translated into English Verse.
Second Edition, Revised. Fcap. 8vo, 6s.

The Works of Horace. Translated into English Verse, with
Life and Notes. 2 vols. New Edition. Crown 8vo, 21s.

Poems and Ballads of Heinrich Heine. Done into English
Verse. Third Edition. Small crown 8vo, 5s.

The Song of the Bell, and other Translations from Schiller,
Goethe, Uhland, and Others. Crown 8vo, 7s. 6d.

Madonna Pia : A Tragedy ; and Three Other Dramas. Crown
8vo, 7s. 6d.

MARTIN.

Catullus. With Life and Notes. Second Edition, Revised and Corrected. Post 8vo, 7s. 6d.

The 'Vita Nuova' of Dante. Translated, with an Introduction and Notes. Third Edition. Small crown 8vo, 5s.

Aladdin : A Dramatic Poem. By ADAM OEHLENSCHLAEGER. Fcap. 8vo, 5s.

Correggio : A Tragedy. By OEHLENSCHLAEGER. With Notes. Fcap. 8vo, 3s.

MARTIN. On some of Shakespeare's Female Characters. By HELENA FAUCIT, Lady MARTIN. Dedicated by permission to Her Most Gracious Majesty the Queen. Fifth Edition. With a Portrait by Lehmann. Demy 8vo, 7s. 6d.

MARWICK. Observations on the Law and Practice in regard to Municipal Elections and the Conduct of the Business of Town Councils and Commissioners of Police in Scotland. By Sir JAMES D. MARWICK, LL.D., Town-Clerk of Glasgow. Royal 8vo, 30s.

MATHESON.

Can the Old Faith Live with the New ? or, The Problem of Evolution and Revelation. By the Rev. GEORGE MATHESON, D.D. Third Edition. Crown 8vo, 7s. 6d.

The Psalmist and the Scientist ; or, Modern Value of the Religious Sentiment. New and Cheaper Edition. Crown 8vo, 5s.

Spiritual Development of St Paul. Third Edition. Cr. 8vo, 5s.

The Distinctive Messages of the Old Religions. Second Edition. Crown 8vo, 5s.

Sacred Songs. New and Cheaper Edition. Crown 8vo, 2s. 6d.

MAURICE. The Balance of Military Power in Europe. An Examination of the War Resources of Great Britain and the Continental States. By Colonel MAURICE, R.A., Professor of Military Art and History at the Royal Staff College. Crown 8vo, with a Map, 6s.

MAXWELL.

Life and Times of the Rt. Hon. William Henry Smith, M.P. By Sir HERBERT MAXWELL, Bart., M.P., F.S.A., &c., Author of 'Passages in the Life of Sir Lucian Elphin.' With Portraits and numerous Illustrations by Herbert Railton, G. L. Seymour, and Others. 2 vols. demy 8vo, 25s. POPULAR EDITION. With a Portrait and other Illustrations. Crown 8vo, 3s. 6d.

Scottish Land Names : Their Origin and Meaning. Being the Rhind Lectures in Archæology for 1893. Post 8vo, 6s.

Meridiana : Noontide Essays. Post 8vo, 7s. 6d.

MELDRUM. The Story of Margrédel : Being a Fireside History of a Fifeshire Family. By D. STORRAR MELDRUM. Cheap Edition. Crown 8vo, 3s. 6d.

MICHEL. A Critical Inquiry into the Scottish Language. With the view of Illustrating the Rise and Progress of Civilisation in Scotland. By FRANCISQUE-MICHEL, F.S.A. Lond. and Scot., Correspondant de l'Institut de France, &c. 4to, printed on hand-made paper, and bound in roxburghe, 66s.

MICHIE.

The Larch : Being a Practical Treatise on its Culture and General Management. By CHRISTOPHER Y. MICHIE, Forester, Cullen House. Crown 8vo, with Illustrations. New and Cheaper Edition. Enlarged, 5s.

The Practice of Forestry. Crown 8vo, with Illustrations. 6s.

MIDDLETON. The Story of Alastair Bhan Comyn ; or, The Tragedy of Dunphail. A Tale of Tradition and Romance. By the Lady MIDDLETON. Square 8vo, 10s. Cheaper Edition, 5s.

MILLER. Landscape Geology. A Plea for the Study of Geology by Landscape Painters. By HUGH MILLER, of H.M. Geological Survey. Crown 8vo, 3s. Cheap Edition, paper cover, 1s.

MINTO.

A Manual of English Prose Literature, Biographical and
Critical: designed mainly to show Characteristics of Style. By W. MINTO,
M.A., Hon. LL.D. of St Andrews; Professor of Logic in the University of Aber-
deen. Third Edition, Revised. Crown 8vo, 7s. 6d.

Characteristics of English Poets, from Chaucer to Shirley.
New Edition, Revised. Crown 8vo, 7s. 6d.

Plain Principles of Prose Composition. Crown 8vo, 1s. 6d.

The Literature of the Georgian Era. Edited, with a Biographical Introduction, by Professor KNIGHT, St Andrews. [*In the press.*

MOIR. Life of Mansie Wauch, Tailor in Dalkeith. By D. M.
MOIR. With 8 Illustrations on Steel, by the late GEORGE CRUIKSHANK. Crown
8vo, 3s. 6d. Another Edition, fcap. 8vo, 1s. 6d.

MOMERIE.

Defects of Modern Christianity, and other Sermons. By
ALFRED WILLIAMS MOMERIE, M.A., D.Sc., LL.D. Fifth Edition. Crown
8vo, 5s.

The Basis of Religion. Being an Examination of Natural
Religion. Third Edition. Crown 8vo, 2s. 6d.

The Origin of Evil, and other Sermons. Seventh Edition,
Enlarged. Crown 8vo, 5s.

Personality. The Beginning and End of Metaphysics, and
a Necessary Assumption in all Positive Philosophy. Fourth Edition, Revised.
Crown 8vo, 3s.

Agnosticism. Fourth Edition, Revised. Crown 8vo, 5s.

Preaching and Hearing ; and other Sermons. Third Edition,
Enlarged. Crown 8vo, 5s.

Belief in God. Third Edition. Crown 8vo, 3s.

Inspiration ; and other Sermons. Second Edition, Enlarged.
Crown 8vo, 5s.

Church and Creed. Third Edition. Crown 8vo, 4s. 6d.

The Future of Religion, and other Essays. Second Edition.
Crown 8vo, 3s. 6d.

MONTAGUE. Military Topography. Illustrated by Practical
Examples of a Practical Subject. By Major-General W. E. MONTAGUE, C.B.,
P.S.C., late Garrison Instructor Intelligence Department, Author of ' Campaign-
ing in South Africa.' With Forty-one Diagrams. Crown 8vo, 5s.

MONTALEMBERT. Memoir of Count de Montalembert. A
Chapter of Recent French History. By Mrs OLIPHANT, Author of the ' Life of
Edward Irving,' &c. 2 vols. crown 8vo, £1, 4s.

MORISON.

Doorside Ditties. By JEANIE MORISON. With a Frontispiece. Crown 8vo, 3s. 6d.

Æolus. A Romance in Lyrics. Crown 8vo, 3s.

There as Here. Crown 8vo, 3s.
₊ *A limited impression on hand-made paper, bound in vellum, 7s. 6d.*

Selections from Poems. Crown 8vo, 4s. 6d.

Sordello. An Outline Analysis of Mr Browning's Poem.
Crown 8vo, 3s.

Of " Fifine at the Fair," " Christmas Eve and Easter Day,"
and other of Mr Browning's Poems. Crown 8vo, 3s.

The Purpose of the Ages. Crown 8vo, 9s.

Gordon : An Our-day Idyll. Crown 8vo, 3s.

Saint Isadora, and other Poems. Crown 8vo, 1s. 6d.

MORISON.
Snatches of Song. Paper, 1s. 6d. ; Cloth, 3s.
Pontius Pilate. Paper, 1s. 6d. ; Cloth, 3s.
Mill o' Forres. Crown 8vo, 1s.
Ane Booke of Ballades. Fcap. 4to, 1s.

MOZLEY. Essays from 'Blackwood.' By the late ANNE
MOZLEY, Author of 'Essays on Social Subjects'; Editor of 'The Letters and
Correspondence of Cardinal Newman,' 'Letters of the Rev. J. B. Mozley,' &c.
With a Memoir by her Sister, FANNY MOZLEY. Post 8vo, 7s. 6d.

MUNRO. On Valuation of Property. By WILLIAM MUNRO,
M.A., Her Majesty's Assessor of Railways and Canals for Scotland. Second
Edition, Revised and Enlarged. 8vo, 3s. 6d.

MURDOCH. Manual of the Law of Insolvency and Bankruptcy:
Comprehending a Summary of the Law of Insolvency, Notour Bankruptcy,
Composition - contracts, Trust - deeds, Cessios, and Sequestrations; and the
Winding-up of Joint-Stock Companies in Scotland ; with Annotations on the
various Insolvency and Bankruptcy Statutes ; and with Forms of Procedure
applicable to these Subjects. By JAMES MURDOCH, Member of the Faculty of
Procurators in Glasgow. Fifth Edition, Revised and Enlarged. 8vo, 12s. net.

MY TRIVIAL LIFE AND MISFORTUNE : A Gossip with
no Plot in Particular. By A PLAIN WOMAN. Cheap Edition. Crown 8vo, 3s. 6d.
By the SAME AUTHOR.
POOR NELLIE. Cheap Edition. Crown 8vo, 3s. 6d.

NAPIER. The Construction of the Wonderful Canon of Loga-
rithms. By JOHN NAPIER of Merchiston. Translated, with Notes, and a
Catalogue of Napier's Works, by WILLIAM RAE MACDONALD. Small 4to, 15s.
A few large-paper copies on Whatman paper, 30s.

NEAVES.
Songs and Verses, Social and Scientific. By An Old Con-
tributor to 'Maga.' By the Hon. Lord NEAVES. Fifth Edition. Fcap. 8vo, 4s.
The Greek Anthology. Being Vol. XX. of 'Ancient Classics
for English Readers.' Crown 8vo, 2s. 6d.

NICHOLSON.
A Manual of Zoology, for the use of Students. With a
General Introduction on the Principles of Zoology. By HENRY ALLEYNE
NICHOLSON, M.D., D.Sc., F.L.S., F.G.S., Regius Professor of Natural History in
the University of Aberdeen. Seventh Edition, Rewritten and Enlarged. Post
8vo, pp. 956, with 555 Engravings on Wood, 18s.
Text-Book of Zoology, for Junior Students. Fifth Edition,
Rewritten and Enlarged. Crown 8vo, with 358 Engravings on Wood, 10s. 6d.
Introductory Text-Book of Zoology, for the use of Junior
Classes. Sixth Edition, Revised and Enlarged, with 166 Engravings, 3s.
Outlines of Natural History, for Beginners : being Descrip-
tions of a Progressive Series of Zoological Types. Third Edition, with
Engravings, 1s. 6d.
A Manual of Palæontology, for the use of Students. With a
General Introduction on the Principles of Palæontology. By Professor H.
ALLEYNE NICHOLSON and RICHARD LYDEKKER, B.A. Third Edition, entirely
Rewritten and greatly Enlarged. 2 vols. 8vo, £3, 3s.
The Ancient Life-History of the Earth. An Outline of the
Principles and Leading Facts of Palæontological Science. Crown 8vo, with 276
Engravings, 10s. 6d.
On the "Tabulate Corals" of the Palæozoic Period, with
Critical Descriptions of Illustrative Species. Illustrated with 15 Lithographed
Plates and numerous Engravings. Super-royal 8vo, 21s.

NICHOLSON.
Synopsis of the Classification of the Animal Kingdom. 8vo, with 106 Illustrations, 6s.
On the Structure and Affinities of the Genus Monticulipora and its Sub-Genera, with Critical Descriptions of Illustrative Species. Illustrated with numerous Engravings on Wood and Lithographed Plates. Super-royal 8vo, 18s.

NICHOLSON.
Communion with Heaven, and other Sermons. By the late MAXWELL NICHOLSON, D.D., St Stephen's, Edinburgh. Crown 8vo, 5s. 6d.
Rest in Jesus. Sixth Edition. Fcap. 8vo, 4s. 6d.

NICHOLSON.
Thoth. A Romance. By JOSEPH SHIELD NICHOLSON, M.A., D.Sc., Professor of Commercial and Political Economy and Mercantile Law in the University of Edinburgh. Third Edition. Crown 8vo, 4s. 6d.
A Dreamer of Dreams. A Modern Romance. Second Edition. Crown 8vo, 6s.

NICOLSON AND MURE. A Handbook to the Local Government (Scotland) Act, 1889. With Introduction, Explanatory Notes, and Index. By J. BADENACH NICOLSON, Advocate, Counsel to the Scotch Education Department, and W. J. MURE, Advocate, Legal Secretary to the Lord Advocate for Scotland. Ninth Reprint. 8vo, 5s.

OLIPHANT.
Masollam : A Problem of the Period. A Novel. By LAURENCE OLIPHANT. 3 vols. post 8vo, 25s. 6d.
Scientific Religion; or, Higher Possibilities of Life and Practice through the Operation of Natural Forces. Second Edition. 8vo, 16s.
Altiora Peto. Cheap Edition. Crown 8vo, boards, 2s. 6d. ; cloth, 3s. 6d. Illustrated Edition. Crown 8vo, cloth, 6s.
Piccadilly. With Illustrations by Richard Doyle. New Edition, 3s. 6d. Cheap Edition, boards, 2s. 6d.
Traits and Travesties ; Social and Political. Post 8vo, 10s. 6d.
Episodes in a Life of Adventure ; or, Moss from a Rolling Stone. Fifth Edition. Post 8vo, 6s.
Haifa : Life in Modern Palestine. Second Edition. 8vo, 7s. 6d.
The Land of Gilead. With Excursions in the Lebanon. With Illustrations and Maps. Demy 8vo, 21s.
Memoir of the Life of Laurence Oliphant, and of Alice Oliphant, his Wife. By Mrs M. O. W. OLIPHANT. Seventh Edition. 2 vols. post 8vo, with Portraits. 21s.
POPULAR EDITION. With a New Preface. Post 8vo, with Portraits. 7s. 6d.

OLIPHANT.
Who was Lost and is Found. By Mrs OLIPHANT. Cr. 8vo, 6s.
Miss Marjoribanks. New Edition. Crown 8vo, 3s. 6d.
The Perpetual Curate, and The Rector. New Edition. Crown 8vo, 3s. 6d.
Salem Chapel, and The Doctor's Family. New Edition. Crown 8vo, 3s. 6d.
Katie Stewart, and other Stories. New Edition. Crown 8vo, cloth, 3s. 6d.
Valentine and his Brother. New Edition. Crown 8vo, 3s. 6d.
Sons and Daughters. Crown 8vo, 3s. 6d.
Katie Stewart. Illustrated boards, 2s. 6d.
Two Stories of the Seen and the Unseen. The Open Door —Old Lady Mary. Paper covers, 1s.

OLIPHANT. Notes of a Pilgrimage to Jerusalem and the Holy
Land. By F. R. OLIPHANT. Crown 8vo, 3s. 6d.

ON SURREY HILLS. By "A SON OF THE MARSHES."
See page 28.

OSWALD. By Fell and Fjord ; or, Scenes and Studies in Ice-
land. By E. J. OSWALD. Post 8vo, with Illustrations. 7s. 6d.

PAGE.
Introductory Text-Book of Geology. By DAVID PAGE, LL.D.,
Professor of Geology in the Durham University of Physical Science, Newcastle,
and Professor LAPWORTH of Mason Science College, Birmingham. With Engravings and Glossarial Index. Twelfth Edition, Revised and Enlarged. 3s. 6d.

Advanced Text-Book of Geology, Descriptive and Industrial.
With Engravings, and Glossary of Scientific Terms. Sixth Edition, Revised and
Enlarged. 7s. 6d.

Introductory Text-Book of Physical Geography. With Sketch-
Maps and Illustrations. Edited by Professor LAPWORTH, LL.D., F.G.S., &c.,
Mason Science College, Birmingham. Thirteenth Edition, Revised and Enlarged,
2s. 6d.

Advanced Text-Book of Physical Geography. Third Edition.
Revised and Enlarged by Professor LAPWORTH. With Engravings. 5s.

PATON.
Spindrift. By Sir J. NOEL PATON. Fcap., cloth, 5s.
Poems by a Painter. Fcap., cloth, 5s.

PATON. Body and Soul. A Romance in Transcendental Path-
ology. By FREDERICK NOEL PATON. Third Edition. Crown 8vo, 1s.

PATRICK. The Apology of Origen in Reply to Celsus. A
Chapter in the History of Apologetics. By the Rev. J. PATRICK, B.D. Post 8vo,
7s. 6d.

PATTERSON.
Essays in History and Art. By R. HOGARTH PATTERSON.
8vo, 12s.

The New Golden Age, and Influence of the Precious Metals
upon the World. 2 vols. 8vo, 31s. 6d.

PAUL. History of the Royal Company of Archers, the Queen's
Body-Guard for Scotland. By JAMES BALFOUR PAUL, Advocate of the Scottish
Bar. Crown 4to, with Portraits and other Illustrations. £2, 2s.

PEILE. Lawn Tennis as a Game of Skill. With latest revised
Laws as played by the Best Clubs. By Captain S. C. F. PEILE, B.S.C. Cheaper
Edition. Fcap., cloth, 1s.

PETTIGREW. The Handy Book of Bees, and their Profitable
Management. By A. PETTIGREW. Fifth Edition, Enlarged, with Engravings.
Crown 8vo, 3s. 6d.

PFLEIDERER. Philosophy and Development of Religion.
Being the Edinburgh Gifford Lectures for 1894. By OTTO PFLEIDERER, D.D.,
Professor of Theology at Berlin University. In 2 vols. post 8vo, 15s. net.

PHILOSOPHICAL CLASSICS FOR ENGLISH READERS.
Edited by WILLIAM KNIGHT, LL.D., Professor of Moral Philosophy, University
of St Andrews. In crown 8vo volumes, with Portraits, price 3s. 6d.
[For List of Volumes, see page 2.

POLLARD. A Study in Municipal Government : The Corpora-
tion of Berlin. By JAMES POLLARD, C.A., Chairman of the Edinburgh Public
Health Committee, and Secretary of the Edinburgh Chamber of Commerce.
Second Edition, Revised. Crown 8vo, 3s. 6d.

POLLOK. The Course of Time : A Poem. By ROBERT POLLOK,
A.M. Cottage Edition, 32mo, 8d. The Same, cloth, gilt edges, 1s. 6d. Another
Edition, with Illustrations by Birket Foster and others, fcap., cloth, 3s. 6d., or
with edges gilt, 4s.

PORT ROYAL LOGIC. Translated from the French; with
Introduction, Notes, and Appendix. By THOMAS SPENCER BAYNES, LL.D., Professor in the University of St Andrews. Tenth Edition, 12mo, 4s.

POTTS AND DARNELL.
Aditus Faciliores : An Easy Latin Construing Book, with
Complete Vocabulary. By A. W. POTTS, M.A., LL.D., and the Rev. C. DARNELL,
M.A., Head-Master of Cargilfield Preparatory School, Edinburgh. Tenth Edition,
fcap. 8vo, 3s. 6d.

Aditus Faciliores Græci. An Easy Greek Construing Book,
with Complete Vocabulary. Fifth Edition, Revised. Feap. 8vo, 3s.

POTTS. School Sermons. By the late ALEXANDER WM. POTTS,
LL.D., First Head-Master of Fettes College. With a Memoir and Portrait.
Crown 8vo, 7s. 6d.

PRINGLE. The Live - Stock of the Farm. By ROBERT O.
PRINGLE. Third Edition. Revised and Edited by JAMES MACDONALD. Crown
8vo, 7s. 6d.

PRYDE. Pleasant Memories of a Busy Life. By DAVID PRYDE,
M.A., LL.D., Author of ' Highways of Literature,' ' Great Men in European History,' ' Biographical Outlines of English Literature,' &c. With a Mezzotint Portrait. Post 8vo, 6s.

PUBLIC GENERAL STATUTES AFFECTING SCOTLAND
from 1707 to 1847, with Chronological Table and Index. 3 vols. large 8vo, £3, 3s.

PUBLIC GENERAL STATUTES AFFECTING SCOTLAND,
COLLECTION OF. Published Annually, with General Index.

RADICAL CURE FOR IRELAND, The. A Letter to the
People of England and Scotland concerning a new Plantation. With 2 Maps.
8vo, 7s. 6d.

RAE. The Syrian Church in India. By GEORGE MILNE RAE,
M.A., D.D., Fellow of the University of Madras; late Professor in the Madras
Christian College. With 6 full-page Illustrations. Post 8vo, 10s. 6d.

RAMSAY. Scotland and Scotsmen in the Eighteenth Century.
Edited from the MSS. of JOHN RAMSAY, Esq. of Ochtertyre, by ALEXANDER
ALLARDYCE, Author of ' Memoir of Admiral Lord Keith, K.B.,' &c. 2 vols.
8vo, 31s. 6d.

RANKIN. The Zambesi Basin and Nyassaland. By DANIEL J.
RANKIN, F.R.S.G.S., M.R.A.S. With 3 Maps and 10 full-page Illustrations.
Post 8vo, 10s. 6d.

RANKIN.
A Handbook of the Church of Scotland. By JAMES RANKIN,
D.D., Minister of Muthill; Author of 'Character Studies in the Old Testament,'
&c. An entirely New and much Enlarged Edition. Crown 8vo, with 2 Maps,
7s. 6d.

The First Saints. Post 8vo, 7s. 6d.

The Creed in Scotland. An Exposition of the Apostles'
Creed. With Extracts from Archbishop Hamilton's Catechism of 1552, John
Calvin's Catechism of 1556, and a Catena of Ancient Latin and other Hymns.
Post 8vo, 7s. 6d.

The Worthy Communicant. A Guide to the Devout Obser-
vance of the Lord's Supper. Limp cloth, 1s. 3d.

The Young Churchman. Lessons on the Creed, the Com-
mandments, the Means of Grace, and the Church. Limp cloth, 1s. 3d.

First Communion Lessons. 24th Edition. Paper Cover, 2d.

RECORDS OF THE TERCENTENARY FESTIVAL OF THE UNIVERSITY OF EDINBURGH. Celebrated in April 1884. Published under the Sanction of the Senatus Academicus. Large 4to, £2, 12s. 6d.

ROBERTSON. The Early Religion of Israel. As set forth by Biblical Writers and Modern Critical Historians. Being the Baird Lecture for 1888-89. By JAMES ROBERTSON, D.D., Professor of Oriental Languages in the University of Glasgow. Fourth Edition. Crown 8vo, 10s. 6d.

ROBERTSON.
Orellana, and other Poems. By J. LOGIE ROBERTSON, M.A. Fcap. 8vo. Printed on hand-made paper. 6s.
A History of English Literature. For Secondary Schools. Crown 8vo, 3s.

ROBERTSON. Our Holiday among the Hills. By JAMES and JANET LOGIE ROBERTSON. Fcap. 8vo, 3s. 6d.

ROBERTSON. Essays and Sermons. By the late W. ROBERT-son, B.D., Minister of the Parish of Sprouston. With a Memoir and Portrait. Crown 8vo, 5s. 6d.

RODGER. Aberdeen Doctors at Home and Abroad. The Story of a Medical School. By ELLA HILL BURTON RODGER. Demy 8vo, 10s. 6d.

ROSCOE. Rambles with a Fishing-Rod. By E. S. ROSCOE. Crown 8vo, 4s. 6d.

ROSS. Old Scottish Regimental Colours. By ANDREW ROSS, S.S.C., Hon. Secretary Old Scottish Regimental Colours Committee. Dedicated by Special Permission to Her Majesty the Queen. Folio. £2, 12s. 6d.

RUTLAND.
Notes of an Irish Tour in 1846. By the DUKE OF RUTLAND, G.C.B. (Lord JOHN MANNERS). New Edition. Crown 8vo, 2s. 6d.

Correspondence between the Right Honble. William Pitt and Charles Duke of Rutland, Lord - Lieutenant of Ireland, 1781-1787. With Introductory Note by JOHN DUKE OF RUTLAND. 8vo, 7s. 6d.

RUTLAND.
Gems of German Poetry. Translated by the DUCHESS OF RUTLAND (Lady JOHN MANNERS). [*New Edition in preparation.*

Impressions of Bad-Homburg. Comprising a Short Account of the Women's Associations of Germany under the Red Cross. Crown 8vo, 1s. 6d.

Some Personal Recollections of the Later Years of the Earl of Beaconsfield, K.G. Sixth Edition. 6d.

Employment of Women in the Public Service. 6d.

Some of the Advantages of Easily Accessible Reading and Recreation Rooms and Free Libraries. With Remarks on Starting and Maintaining them. Second Edition. Crown 8vo, 1s.

A Sequel to Rich Men's Dwellings, and other Occasional Papers. Crown 8vo, 2s. 6d.

Encouraging Experiences of Reading and Recreation Rooms, Aims of Guilds, Nottingham Social Guide, Existing Institutions, &c., &c. Crown 8vo, 1s.

SCHEFFEL. The Trumpeter. A Romance of the Rhine. By JOSEPH VICTOR VON SCHEFFEL. Translated from the Two Hundredth German Edition by JESSIE BECK and LOUISA LORIMER. With an Introduction by Sir THEODORE MARTIN, K.C.B. Long 8vo, 3s. 6d.

SCHILLER. Wallenstein. A Dramatic Poem. By FRIEDRICH VON SCHILLER. Translated by C. G. N. LOCKHART. Fcap. 8vo, 7s. 6d.

SCOTCH LOCH FISHING. By "BLACK PALMER." Crown 8vo,
Interleaved with blank pages, 4s.

SCOUGAL. Prisons and their Inmates; or, Scenes from a
Silent World. By FRANCIS SCOUGAL. Crown 8vo, boards, 2s.

SELLAR'S Manual of the Acts relating to Education in Scotland. By J. EDWARD GRAHAM, B.A. Oxon., Advocate. Ninth Edition. Demy
8vo, 12s. 6d.

SETH.
Scottish Philosophy. A Comparison of the Scottish and
German Answers to Hume. Balfour Philosophical Lectures, University of
Edinburgh. By ANDREW SETH, LL.D., Professor of Logic and Metaphysics in
Edinburgh University. Second Edition. Crown 8vo, 5s.

Hegelianism and Personality. Balfour Philosophical Lectures.
Second Series. Second Edition. Crown 8vo, 5s.

SETH. A Study of Ethical Principles. By JAMES SETH, M.A.,
Professor of Philosophy in Brown University, U.S.A. Post 8vo, 10s. 6d. net.

SHADWELL. The Life of Colin Campbell, Lord Clyde. Illustrated by Extracts from his Diary and Correspondence. By Lieutenant-General
SHADWELL, C.B. With Portrait, Maps, and Plans. 2 vols. 8vo, 36s.

SHAND.
Half a Century; or, Changes in Men and Manners. By
ALEX. INNES SHAND, Author of 'Kilcarra,' 'Against Time,' &c. Second Edition.
8vo, 12s. 6d.

Letters from the West of Ireland. Reprinted from the
'Times.' Crown 8vo, 5s.

SHARPE. Letters from and to Charles Kirkpatrick Sharpe.
Edited by ALEXANDER ALLARDYCE, Author of 'Memoir of Admiral Lord Keith,
K.B.,' &c. With a Memoir by the Rev. W. K. R. BEDFORD. In 2 vols. 8vo.
Illustrated with Etchings and other Engravings. £2, 12s. 6d.

SIM. Margaret Sim's Cookery. With an Introduction by L. B.
WALFORD, Author of 'Mr Smith: A Part of his Life,' &c. Crown 8vo, 5s.

SIMPSON. The Wild Rabbit in a New Aspect; or, Rabbit-
Warrens that Pay. A book for Landowners, Sportsmen, Land Agents, Farmers,
Gamekeepers, and Allotment Holders. A Record of Recent Experiments con-
ducted on the Estate of the Right Hon. the Earl of Wharncliffe at Wortley Hall.
By J. SIMPSON. Small crown 8vo, 5s.

SKELTON.
Maitland of Lethington; and the Scotland of Mary Stuart.
A History. By JOHN SKELTON, Advocate, C.B., LL.D., Author of 'The Essays
of Shirley.' Limited Edition, with Portraits. Demy 8vo, 2 vols., 28s. net.

The Handbook of Public Health. A Complete Edition of the
Public Health and other Sanitary Acts relating to Scotland. Annotated, and
with the Rules, Instructions, and Decisions of the Board of Supervision brought
up to date with relative forms. Second Edition. With Introduction, containing
the Administration of the Public Health Act in Counties. 8vo, 8s. 6d.

The Local Government (Scotland) Act in Relation to Public
Health. A Handy Guide for County and District Councillors, Medical Officers,
Sanitary Inspectors, and Members of Parochial Boards. Second Edition. With
a new Preface on appointment of Sanitary Officers. Crown 8vo, 2s.

SKRINE. Columba: A Drama. By JOHN HUNTLEY SKRINE,
Warden of Glenalmond; Author of 'A Memory of Edward Thring.' Fcap. 4to, 6s.

SMITH. For God and Humanity. A Romance of Mount Carmel.
By HASKETT SMITH, Author of 'The Divine Epiphany,' &c. 3 vols. post 8vo,
25s. 6d.

SMITH.

Thorndale; or, The Conflict of Opinions. By WILLIAM SMITH, Author of 'A Discourse on Ethics,' &c. New Edition. Crown 8vo, 10s. 6d.

Gravenhurst; or, Thoughts on Good and Evil. Second Edition. With Memoir and Portrait of the Author. Crown 8vo, 8s.

The Story of William and Lucy Smith. Edited by GEORGE MERRIAM. Large post 8vo, 12s. 6d.

SMITH. Memoir of the Families of M'Combie and Thoms, originally M'Intosh and M'Thomas. Compiled from History and Tradition. By WILLIAM M'COMBIE SMITH. With Illustrations. 8vo, 7s. 6d.

SMITH. Greek Testament Lessons for Colleges, Schools, and Private Students, consisting chiefly of the Sermon on the Mount and the Parables of our Lord. With Notes and Essays. By the Rev. J. HUNTER SMITH, M.A., King Edward's School, Birmingham. Crown 8vo, 6s.

SMITH. The Secretary for Scotland. Being a Statement of the Powers and Duties of the new Scottish Office. With a Short Historical Introduction, and numerous references to important Administrative Documents. By W. C. SMITH, LL.B., Advocate. 8vo, 6s.

"SON OF THE MARSHES, A."

From Spring to Fall; or, When Life Stirs. By "A SON OF THE MARSHES. Crown 8vo, 3s. 6d.

Within an Hour of London Town: Among Wild Birds and their Haunts. Edited by J. A. OWEN. Cheap Uniform Edition. Crown 8vo, 3s. 6d.

With the Woodlanders, and By the Tide. Cheap Uniform Edition. Crown 8vo, 3s. 6d.

On Surrey Hills. Cheap Uniform Edition. Crown 8vo, 3s. 6d.

Annals of a Fishing Village. Cheap Uniform Edition. Crown 8vo, 3s. 6d.

SORLEY. The Ethics of Naturalism. Being the Shaw Fellowship Lectures, 1884. By W. R. SORLEY, M.A., Fellow of Trinity College, Cambridge, Professor of Logic and Philosophy in University College of South Wales. Crown 8vo, 6s.

SPEEDY. Sport in the Highlands and Lowlands of Scotland with Rod and Gun. By TOM SPEEDY. Second Edition, Revised and Enlarged. With Illustrations by Lieut.-General Hope Crealocke, C.B., C.M.G., and others. 8vo, 15s.

SPROTT. The Worship and Offices of the Church of Scotland. By GEORGE W. SPROTT, D.D., Minister of North Berwick. Crown 8vo, 6s.

STATISTICAL ACCOUNT OF SCOTLAND. Complete, with Index. 15 vols. 8vo, £16, 16s.

STEPHENS.

The Book of the Farm; detailing the Labours of the Farmer, Farm-Steward, Ploughman, Shepherd, Hedger, Farm-Labourer, Field-Worker, and Cattle-man. Illustrated with numerous Portraits of Animals and Engravings of Implements, and Plans of Farm Buildings. Fourth Edition. Revised, and in great part Rewritten by JAMES MACDONALD, F.R.S.E., Secretary, Highland and Agricultural Society of Scotland. Complete in Six Divisional Volumes, bound in cloth, each 10s. 6d., or handsomely bound, in 3 volumes, with leather back and gilt top, £3, 3s.

The Book of Farm Implements and Machines. By J. SLIGHT and R. SCOTT BURN, Engineers. Edited by HENRY STEPHENS. Large 8vo, £2, 2s.

Catechism of Agriculture.   [*New Edition in preparation.*

STEVENSON. British Fungi. (Hymenomycetes.) By Rev.
John Stevenson, Author of 'Mycologia Scotia,' Hon. Sec. Cryptogamic Society of Scotland. Vols. I. and II., post 8vo, with Illustrations, price 12s. 6d. net each.

STEWART.
Advice to Purchasers of Horses. By John Stewart, V.S. New Edition. 2s. 6d.
Stable Economy. A Treatise on the Management of Horses in relation to Stabling, Grooming, Feeding, Watering, and Working. Seventh Edition. Fcap. 8vo, 6s. 6d.

STEWART. A Hebrew Grammar, with the Pronunciation, Syllabic Division and Tone of the Words, and Quantity of the Vowels. By Rev. Duncan Stewart, D.D. Fourth Edition. 8vo, 3s. 6d.

STEWART. Boethius: An Essay. By Hugh Fraser Stewart, M.A., Trinity College, Cambridge. Crown 8vo, 7s. 6d.

STODDART. Sir Philip Sidney: Servant of God. By Anna M. Stoddart. Illustrated by Margaret L. Huggins. With a New Portrait of Sir Philip Sidney. Small 4to, with a specially designed Cover. 5s.

STODDART. Angling Songs. By Thomas Tod Stoddart. New Edition, with a Memoir by Anna M. Stoddart. Crown 8vo, 7s. 6d.

STORMONTH.
Etymological and Pronouncing Dictionary of the English Language. Including a very Copious Selection of Scientific Terms. For use in Schools and Colleges, and as a Book of General Reference. By the Rev. James Stormonth. The Pronunciation carefully revised by the Rev. P. H. Phelp, M.A. Cantab. Eleventh Edition, with Supplement. Crown 8vo, pp. 800. 7s. 6d.
Dictionary of the English Language, Pronouncing, Etymological, and Explanatory. Revised by the Rev. P. H. Phelp. Library Edition. New and Cheaper Edition, with Supplement. Imperial 8vo, handsomely bound in half morocco, 21s.
The School Etymological Dictionary and Word-Book. Fourth Edition. Fcap. 8vo, pp. 254. 2s.

STORY.
Nero; A Historical Play. By W. W. Story, Author of 'Roba di Roma.' Fcap. 8vo, 6s.
Vallombrosa. Post 8vo, 5s.
Poems. 2 vols., 7s. 6d.
Fiammetta. A Summer Idyl. Crown 8vo, 7s. 6d.
Conversations in a Studio. 2 vols. crown 8vo, 12s. 6d.
Excursions in Art and Letters. Crown 8vo, 7s. 6d.
A Poet's Portfolio: Later Readings. 18mo, 3s. 6d.

STURGIS.
John-a-Dreams. A Tale. By Julian Sturgis. New Edition. Crown 8vo, 3s. 6d.
Little Comedies, Old and New. Crown 8vo, 7s. 6d.

SUTHERLAND (DUCHESS OF). How I Spent my Twentieth Year. Being a Record of a Tour Round the World, 1886-87. By the Duchess of Sutherland (Marchioness of Stafford). With Illustrations. Crown 8vo, 7s. 6d.

SUTHERLAND. Handbook of Hardy Herbaceous and Alpine Flowers, for General Garden Decoration. Containing Descriptions of upwards of 1000 Species of Ornamental Hardy Perennial and Alpine Plants; along with Concise and Plain Instructions for their Propagation and Culture. By William Sutherland, Landscape Gardener; formerly Manager of the Herbaceous Department at Kew. Crown 8vo, 7s. 6d.

TAYLOR. The Story of my Life. By the late Colonel
MEADOWS TAYLOR, Author of 'The Confessions of a Thug,' &c., &c. Edited by
his Daughter. New and Cheaper Edition, being the Fourth. Crown 8vo, 6s.

THOLUCK. Hours of Christian Devotion. Translated from
the German of A. Tholuck, D.D., Professor of Theology in the University of
Halle. By the Rev. ROBERT MENZIES, D.D. With a Preface written for this
Translation by the Author. Second Edition. Crown 8vo, 7s. 6d.

THOMSON. South Sea Yarns. By Basil Thomson. With 10
Full-page Illustrations. Crown 8vo, 6s.

The Diversions of a Prime Minister. In 1 vol. With Illus-
trations. Small demy 8vo.   [*In the press.*

THOMSON.
Handy Book of the Flower-Garden : being Practical Direc-
tions for the Propagation, Culture, and Arrangement of Plants in Flower-
Gardens all the year round. With Engraved Plans. By DAVID THOMSON,
Gardener to his Grace the Duke of Buccleuch, K.T., at Drumlanrig. Fourth
and Cheaper Edition. Crown 8vo, 5s.

The Handy Book of Fruit-Culture under Glass: being a
series of Elaborate Practical Treatises on the Cultivation and Forcing of Pines,
Vines, Peaches, Figs, Melons, Strawberries, and Cucumbers. With Engravings
of Hothouses, &c. Second Edition, Revised and Enlarged. Crown 8vo, 7s. 6d.

THOMSON. A Practical Treatise on the Cultivation of the
Grape Vine. By WILLIAM THOMSON, Tweed Vineyards. Tenth Edition. 8vo, 5s.

THOMSON. Cookery for the Sick and Convalescent. With
Directions for the Preparation of Poultices, Fomentations, &c. By BARBARA
THOMSON. Feap. 8vo, 1s. 6d.

THORBURN. Asiatic Neighbours. By S. S. THORBURN, Bengal
Civil Service, Author of 'Bannú; or, Our Afghan Frontier,' 'Musalmans and
Money-Lenders in the Punjab.' With Four Maps. Demy 8vo, 12s. net.

THORNTON. Opposites. A Series of Essays on the Unpopular
Sides of Popular Questions. By LEWIS THORNTON. 8vo, 12s. 6d.

TOM CRINGLE'S LOG. A New Edition, with Illustrations.
Crown 8vo, cloth gilt, 5s. Cheap Edition, 2s.

TRANSACTIONS OF THE HIGHLAND AND AGRICUL-
TURAL SOCIETY OF SCOTLAND. Published annually, price 5s.

TRAVEL, ADVENTURE, AND SPORT. From 'Blackwood's
Magazine.' Uniform with 'Tales from Blackwood.' In 12 Parts, each price 1s.
Handsomely bound in 6 vols., cloth, 15s.; half calf, 25s.

TRAVERS. Mona Maclean, Medical Student. A Novel. By
GRAHAM TRAVERS. Ninth Edition. Crown 8vo, 6s.

TULLOCH.
Rational Theology and Christian Philosophy in England in
the Seventeenth Century. By JOHN TULLOCH, D.D., Principal of St Mary's College in the University of St Andrews; and one of her Majesty's Chaplains in
Ordinary in Scotland. Second Edition. 2 vols. 8vo, 16s.

Modern Theories in Philosophy and Religion. 8vo, 15s.

Luther, and other Leaders of the Reformation. Third Edition, Enlarged. Crown 8vo, 3s. 6d.

Memoir of Principal Tulloch, D.D., LL.D. By Mrs OLIPHANT,
Author of 'Life of Edward Irving.' Third and Cheaper Edition. 8vo, with
Portrait, 7s. 6d.

TWEEDIE. The Arabian Horse: His Country and People.
By Major-General W. TWEEDIE, C.S.I., Bengal Staff Corps; for many years
H.B.M.'s Consul-General, Baghdad, and Political Resident for the Government
of India in Turkish Arabia. In one vol. royal 4to, with Seven Coloured Plates
and other Illustrations, and a Map of the Country. Price £3, 3s. net.

VEITCH.
The History and Poetry of the Scottish Border : their Main
Features and Relations. By JOHN VEITCH, LL.D., Professor of Logic and
Rhetoric in the University of Glasgow. New and Enlarged Edition. 2 vols.
demy 8vo, 16s.

Institutes of Logic. Post 8vo, 12s. 6d.

The Feeling for Nature in Scottish Poetry. From the Ear-
liest Times to the Present Day. 2 vols. feap. 8vo, in roxburghe binding, 15s.

Merlin and other Poems. Fcap. 8vo, 4s. 6d.

Knowing and Being. Essays in Philosophy. First Series.
Crown 8vo, 5s.

VIRGIL. The Æneid of Virgil. Translated in English Blank
Verse by G. K. RICKARDS, M.A., and Lord RAVENSWORTH. 2 vols. fcap. 8vo, 10s.

WACE. The Christian Faith and Recent Agnostic Attacks.
By the Rev. HENRY WACE, D.D., Principal of King's College, London ; Preacher
of Lincoln's Inn ; Chaplain to the Queen. In one vol. post 8vo. [*In preparation.*

WADDELL. An Old Kirk Chronicle : Being a History of Auld-
hame, Tyninghame, and Whitekirk, in East Lothian. From Session Records,
1615 to 1850. By Rev. P. HATELY WADDELL, B.D., Minister of the United
Parish. Small Paper Edition, 200 Copies. Price £1. Large Paper Edition, 50
Copies. Price £1, 10s.

WALFORD. Four Biographies from 'Blackwood' : Jane Taylor,
Hannah More, Elizabeth Fry, Mary Somerville. By L. B. WALFORD. Crown
8vo, 5s.

WALKER. The Teaching of Jesus in His Own Words. By the
Rev. JOHN C. WALKER. Crown 8vo, 3s. 6d.

WARREN'S (SAMUEL) WORKS :—
Diary of a Late Physician. Cloth, 2s. 6d. ; boards, 2s.

Ten Thousand A-Year. Cloth, 3s. 6d. ; boards, 2s. 6d.

Now and Then. The Lily and the Bee. Intellectual and
Moral Development of the Present Age. 4s. 6d.

Essays : Critical, Imaginative, and Juridical. 5s.

WEBSTER. The Angler and the Loop - Rod. By DAVID
WEBSTER. Crown 8vo, with Illustrations, 7s. 6d.

WENLEY.
Socrates and Christ : A Study in the Philosophy of Religion.
By R. M. WENLEY, M.A., D.Sc., Lecturer on Mental and Moral Philosophy in
Queen Margaret College, Glasgow ; formerly Examiner in Philosophy in the
University of Glasgow. Crown 8vo, 6s.

Aspects of Pessimism. Crown 8vo, 6s.

WERNER. A Visit to Stanley's Rear-Guard at Major Bartte-
lot's Camp on the Aruhwimi. With an Account of River-Life on the Congo.
By J. R. WERNER, F.R.G.S., Engineer, late in the Service of the Etat Indepen-
dant du Congo. With Maps, Portraits, and other Illustrations. 8vo, 16s.

WESTMINSTER ASSEMBLY. Minutes of the Westminster
Assembly, while engaged in preparing their Directory for Church Government,
Confession of Faith, and Catechisms (November 1644 to March 1649). Edited
by the Rev. Professor ALEX. T. MITCHELL, of St Andrews, and the Rev. JOHN
STRUTHERS, LL.D. With a Historical and Critical Introduction by Professor
Mitchell. 8vo, 15s.

WHITE.
The Eighteen Christian Centuries. By the Rev. JAMES
WHITE. Seventh Edition. Post 8vo, with Index, 6s.

History of France, from the Earliest Times. Sixth Thousand.
Post 8vo, with Index, 6s.

WHITE.
Archæological Sketches in Scotland—Kintyre and Knapdale. By Colonel T. P. WHITE, R.E., of the Ordnance Survey. With numerous Illustrations. 2 vols. folio, £4, 4s. Vol. I., Kintyre, sold separately, £2, 2s.

The Ordnance Survey of the United Kingdom. A Popular Account. Crown 8vo, 5s.

WILLIAMSON.
The Horticultural Exhibitor's Handbook. A Treatise on Cultivating, Exhibiting, and Judging Plants, Flowers, Fruits, and Vegetables. By W. WILLIAMSON, Gardener. Revised by MALCOLM DUNN, Gardener to his Grace the Duke of Buccleuch and Queensberry, Dalkeith Park. Crown 8vo, 3s. 6d.

WILLIAMSON.
Poems of Nature and Life. By DAVID R. WILLIAMSON, Minister of Kirkmaiden. Fcap. 8vo, 3s.

WILLIAMSON.
Light from Eastern Lands on the Lives of Abraham, Joseph, and Moses. By the Rev. ALEX. WILLIAMSON, Author of 'The Missionary Heroes of the Pacific,' 'Sure and Comfortable Words,' 'Ask and Receive,' &c. Crown 8vo, 3s. 6d.

WILLS AND GREENE.
Drawing-Room Dramas for Children. By W. G. WILLS and the Hon. Mrs GREENE. Crown 8vo, 6s.

WILSON.
Works of Professor Wilson. Edited by his Son-in-Law, Professor FERRIER. 12 vols. crown 8vo, £2, 8s.

Christopher in his Sporting-Jacket. 2 vols., 8s.

Isle of Palms, City of the Plague, and other Poems. 4s.

Lights and Shadows of Scottish Life, and other Tales. 4s.

Essays, Critical and Imaginative. 4 vols., 16s.

The Noctes Ambrosianæ. 4 vols., 16s.

Homer and his Translators, and the Greek Drama. Crown 8vo, 4s.

WITHIN AN HOUR OF LONDON TOWN.
Among Wild Birds and their Haunts. By "A SON OF THE MARSHES." *See page 28.*

WITH THE WOODLANDERS, AND BY THE TIDE.
By "A SON OF THE MARSHES." *See page 28.*

WORSLEY.
Poems and Translations. By PHILIP STANHOPE WORSLEY, M.A. Edited by EDWARD WORSLEY. Second Edition, Enlarged. Fcap. 8vo, 6s.

Homer's Odyssey. Translated into English Verse in Spen- serian Stanza. By P. S. Worsley. Third Edition. 2 vols. fcap., 12s.

Homer's Iliad. Translated by P. S. Worsley and Prof. Con- ington. 2 vols. crown 8vo, 21s.

YATE.
England and Russia Face to Face in Asia. A Record of Travel with the Afghan Boundary Commission. By Captain A. C. YATE, Bombay Staff Corps. 8vo, with Maps and Illustrations, 21s.

YATE.
Northern Afghanistan; or, Letters from the Afghan Boundary Commission. By Major C. E. YATE, C.S.I., C.M.G. Bombay Staff Corps, F.R.G.S. 8vo, with Maps, 18s.

YOUNG.
A Story of Active Service in Foreign Lands. Com- piled from Letters sent home from South Africa, India, and China, 1856-1882. By Surgeon-General A. GRAHAM YOUNG, Author of 'Crimean Cracks.' Crown 8vo, Illustrated, 7s. 6d.

YULE.
Fortification : For the use of Officers in the Army, and Readers of Military History. By Colonel YULE, Bengal Engineers. 8vo, with Numerous Illustrations, 10s.

10/94.

www.ingramcontent.com/pod-product-compliance
Lightning Source LLC
Chambersburg PA
CBHW030905270326
41929CB00008B/582